THE EOCENE-OLIGOCENE TRANSITION

PARADISE LOST

Critical Moments in Paleobiology and Earth History Series

David J. Bottjer and Richard K. Bambach, Editors

Critical Moments in Paleobiology and Earth History Series
David J. Bottjer and Richard K. Bambach, Editors

Mark A. S. McMenamin and Dianna Schulte McMenamin,
   The Emergence of Animals: The Cambrian Breakthrough

Douglas H. Erwin, *The Great Paleozoic Crisis:*
   Life and Death in the Permian

Betsey Dexter Dyer and Robert Alan Obar, *Tracing the*
   History of Eukaryotic Cells: The Enigmatic Smile

Donald R. Prothero,
   The Eocene-Oligocene Transition: Paradise Lost

Perspectives in Paleobiology and Earth History Series
David J. Bottjer and Richard K. Bambach, Editors

Anthony Hallam,
   Phanerozoic Sea-Level Changes

Columbia University Press Advisory Committee for
Paleontology:

DAVID J. BOTTJER, CHAIR
RICHARD K. BAMBACH
DAVID L. DILCHER
NILES ELDREDGE
S. DAVID WEBB

The concept for these series was suggested by Mark and
Dianna McMenamin, whose book, *The Emergence of Animals*,
was the first to be published.

The Eocene-Oligocene Transition
Paradise Lost

DONALD R. PROTHERO

COLUMBIA UNIVERSITY PRESS
NEW YORK

The typesetting of this book was done by the author
as a special project of the science publishing program
of Columbia University Press.

Columbia University Press
New York   Chichester, West Sussex
Copyright © 1994 Columbia University Press
All rights reserved

Library of Congress Cataloging-in-Publication Data

Prothero, Donald R.
The Eocene-Oligocene transition: paradise lost/
Donald R. Prothero.
p. cm. —(Critical moments in paleobiology and earth history
series)
Includes bibliographical references and index.
ISBN 0–231–08090–5. —ISBN 0–231–08091–3 (paper)
1. Eocene-Oligocene boundary. 2. Extinction (Biology)
3. Paleontology—Eocene. 4. Paleontology—Oligocene. I. Title. II. Series.
QE692.8.P76 1994
560'.178—dc20   93-44535 CIP
∞

Casebound editions of Columbia University Press books are printed on
permanent and durable acid free paper.

Printed in the United States of America
p 10 9 8 7 6 5 4 3 2 1

Spectacular pinnacles of the early Oligocene Brule Formation, Badlands National Park, South Dakota, as photographed by the pioneer geologist N.H. Darton in 1898. Floodplain mudstones capped by cross-bedded channel sandstones form the balanced rocks and hoodoos that rapidly erode away. The ancient soil horizons in these sediments record a much drier woodland-savanna habitat developed after the early Oligocene climatic deterioration. (Photograph courtesy U.S. Geological Survey).

To My Parents

Clifford and Shirley Prothero
*for their years of love and support
and for never insisting that
I pursue a "practical" career*

Artist's conception of a family of brontotheres weathering a storm. These elephant-size beasts were close relatives of horses, tapirs, and rhinos, but lived in North America and Asia during the middle and late Eocene. The last of the brontotheres had huge, blunt, paired horns and died out in the extinctions at the beginning of the Oligocene. (Painting by Z. Burian).

O unexpected stroke, worse than of Death!
Must I thus leave thee Paradise? thus leave
Thee native soil, these happy walks and shades,
Fit haunt of the Gods? where I had hope to spend,
Quiet though sad, the respite of that day
That must be mortal to us both. O flow'rs
That never will in other climate grow,
My early visitation, and my last
At ev'n, which I bred up with tender hand
From the first op'ning bud, and gave ye names,
Who now shall rear yee to the Sun, or rank
Your tribes, and water from th' ambrosial fount?
Thee lastly nuptial bower, by mee adorn'd
With what to sight or smell was sweet; from thee
How shall I part, and whither wander down
Into a lower world, to this obscure
And wild, how shall we breathe in other air
Less pure, accustom'd to immortal fruits?
—JOHN MILTON, *PARADISE LOST* (1667)

CONTENTS

PREFACE

Nearly everyone has puzzled over the extinction of the dinosaurs. Scores of books and thousands of scientific papers have appeared on the Cretaceous-Tertiary extinctions, including a cover story in *Time* magazine. The extinction of large mammals 10,000 years ago at the end of the last Ice Age is another popular topic. Vanishing mammoths, mastodonts, sabertoothed cats, and ground sloths fascinate not only the lay public and professional paleontologists but also glacial geologists, paleoclimatologists, and archeologists. Between these two crises, however, was an episode that has been comparatively neglected. Before 38 million years ago, the earth was still in its warm, subtropical "greenhouse" state that had prevailed since early in the age of dinosaurs. By 33 million years ago, there were glaciers on Antarctica, and the planet had begun its shift to the modern "icehouse" state. This transition, spanning ten million years from the middle Eocene to the late Oligocene, is fundamental to the history of life since the demise of the dinosaurs. In essence, the Eocene "paradise" was lost, and the modern world began with the Oligocene.

Yet compared to the Cretaceous and Pleistocene extinctions, little interest has been focused on the Eocene-Oligocene transition. This is partly because, until recently, we had only vague notions as to when events recorded in strata around the world actually happened. The decade of the 1980s, however, revolutionized our understanding of this important time interval. Many new techniques, from magnetostratigraphy to ^{40}Ar/^{39}Ar dating to isotopic analysis of deep-sea sediments, were applied intensively to marine and terrestrial rocks spanning this transition. By the end of the decade, deep-sea drilling produced many new, complete records of the history of the oceans during the Eocene and Oligocene, and drilling and seismic reflections revealed the presence of Oligocene glaciers in Antarctica. Detailed studies of nonmarine sediments allowed the first precise dating and correlation of terrestrial history with the marine record. Terrestrial soils, plants, pollen, land snails, reptiles and amphibians, and especially mammals were reviewed and restudied—many for the first time in fifty years. In short, there has been an explosion of research on the Eocene-Oligocene transition, and most of what textbooks and nonspecialists think about it is either badly out of date, or just plain wrong.

By the late 1980s, a number of paleontologists felt that this situation should be corrected. On August 1–5, 1989, Bill Berggren, Phil Bjork, and I organized a Penrose Conference of sixty-one international experts to update everyone on the latest developments. The Penrose Conference was followed three months later by a theme session at the annual meeting of the Geological Society of America in St. Louis on November 6, 1989, where we presented our conclusions to the geological community. Although much of this new information is now available, most of it is written at a highly technical level. This book is intended to give nonspecialists a quick summary of our current understanding of this fascinating and critical moment in the development of the modern world.

ACKNOWLEDGMENTS

This project would not have been possible without the encouragement and support of many people over the years. My initial research in the White River Group built on decades of collecting and stratigraphic work by the late Morris F. Skinner. Bob Emry taught me the details of White River stratigraphy and mammals. Malcolm C. McKenna, my graduate adviser, introduced me to the Eocene and encouraged me to combine new techniques, such as magnetic stratigraphy, to solve problems where these methods had never been applied before. Earl Manning taught me a lot about mammalian systematics and biostratigraphy. Neil Opdyke and Dennis Kent inducted me into the mysteries of "paleomagic," and I learned much about micropaleontology and marine geology from Dave Lazarus and Jim Hays.

ACKNOWLEDGMENTS

I have received of expert advice on nearly every Eocene and Oligocene outcrop in western North America. I cannot possibly acknowledge here all the people who have helped me. However, I am particularly grateful to John Armentrout, Phil Bjork, Tom Deméré, Emmett Evanoff, the late Edwin Galbreath, Alden Hamblin, Tom Kelly, Jay Lillegraven, Spencer Lucas, Mark Mason, the late Gregg Ostrander, Mark Roeder, Margaret Stevens, John Storer, Rich Stucky, Jim Swinehart, Richard Tedford, Alan Tabrum, Mike Voorhies, Steve Walsh, and Jack Wilson for their guidance in the field. I have benefited from the hard work and good company of many undergraduate field assistants over the years, including Jeff Amato, Susan Briggs, Jill Bush, Erin Campbell, Joby Campbell, Dani Crosby, Jennifer Chean, Huxley Dozier, Priscilla Duskin, Jon Erskine, Jim Finegan, John Foster, Jon Frenzel, Dana Gilchrist, Karen Gonzalez, Kecia Harris, Steve King, Allison Kozak, Rob Lander, Walter Lohr, Dave Lundquist, Heidi Shlosar, Tim Tierney, Annie Walton, and Erin Wilson. I thank Chuck Denham, Dennis Kent, Bill Roggenthen, and Joe Kirschvink for access to their paleomagnetics laboratories. Carl Swisher has collaborated with me on dating many key areas, which has immensely improved our calibration of the Eocene and Oligocene.

The Penrose Conference and GSA theme session would never have occurred without the initial suggestions of Vince Santucci, or the hard work and enthusiasm of my co-conveners, Bill Berggren and Phil Bjork. I am particularly grateful to Bill Berggren for all his help and advice in preparing our symposium volume. I thank Bill Schopf and the Center for the Study of the Evolution and Origin of Life at the University of California at Los Angeles for financial support of foreign scientists at the Penrose Conference. My own research over the years was supported by the Petroleum Research Fund of the American Chemical Society, a Guggenheim Fellowship, and by grants EAR87–08221 and EAR91–17819 from the National Science Foundation.

This book was produced entirely by the author as camera-ready copy on a MacIntosh IIsi computer, and printed on a 1000 dpi laser printer. I am grateful to Bill Berggren, Herb Brauer, Jim Kennett, Spencer Lucas, Brian McGowran, Greg

ACKNOWLEDGMENTS

Retallack, Robert Schoch, Scott Wing, and Jim Zachos for reading an earlier draft of this manuscript and pointing out problems. I thank Ed Lugenbeel at Columbia University Press for encouraging me to write this book, Connie Barlow for carefully and thoughtfully copyediting the manuscript, and Teresa Bonner and Laura Wood for production assistance. My father, Clifford R. Prothero, did most of the page layouts and darkroom work.

Finally, I thank my parents for their support and guidance, even when I got hooked on paleontology as a boy and seemed to have no hope of pursuing a practical career. My father taught me a great deal about art, photography, and book production, and this training has been invaluable in producing my publications. I thank my wife Anita for putting up with my moods and absences, as I disappeared into the computer room for weeks while I wrote this book. Although I "gave birth" to four books during my sabbatical in 1991–1992, nothing I write will ever compare to her bringing our son Erik into the world.

FIGURE 1.1. Panorama of the Bighorn Basin, Wyoming, from the top of Polecat Bench, showing the badlands of the lower Eocene Willwood Formation. These drab gray mudstones were deposited in humid tropical conditions, and yield thousands of teeth and jaws (and occasionally skeletons) of the mammals which inhabited Eocene jungles.

Greenhouse of the Dinosaurs
The Early Eocene

*The great mass of gypsum may be considered as a purely
fresh-water deposit, containing land and fluviatile shells,
together with fragments of palm-wood, and great
numbers of skeletons of quadrupeds and birds, an
assemblage of organic remains which has given great
celebrity to the Paris Basin. The bones of fresh-water fish,
and of crocodiles, and many land and fluviatile reptiles
occur in this rock . . . . The heat of European latitudes
during the Eocene period . . . seem[s] . . . equal to that now
experienced between the tropics.*

—CHARLES LYELL, *PRINCIPLES OF GEOLOGY* (1833)

Driving through the Bighorn Basin of Wyoming, you pass miles and miles of badlands and sagebrush scrub (figure 1.1). The region is too arid for farming, except in irrigated valleys of the Bighorn and Shoshone Rivers. In a few places grass and scrub is sufficient for cattle grazing, but the badlands are of no agricultural value. Near the Montana border are a number of oil wells, but most of the Bighorn Basin is not rich in minerals. Summer days regularly blaze above 32°C (90°F), and during the winter, howling blizzards whip the snow along, driving the wind chill down to -45°C (-50°F). No wonder most tourists hurry through on their way west to Yellowstone National Park. If they stop at all, it is for the Buffalo Bill Museum in Cody, or the fake Indian trinkets at the many tacky tourist traps that catch their attention.

GREENHOUSE OF THE DINOSAURS

Paleontologists and geologists have a different view of the world. Where others see dry and barren badlands, paleontologists see fossil beds teeming with ancient life. Where others see endless rocks and pebbles, paleontologists scan the ground and find fossil bones and teeth. Where others curse the shales that cause roads to slide, paleobotanists find fossil leaves. Where others see red, gray, and green color bands on the buttes, paleopedologists see evidence of ancient soil horizons that reveal past climates. Where others see odd-looking sandstone boulders, sedimentologists see ancient rivers meandering across a tropical floodplain. In the laboratory, geochemists extract evidence of ancient atmospheric conditions from fossil teeth. In short, the barren landscape of the Bighorn Basin may look hostile and uninteresting to the people who zoom by at 65 miles per hour, but the geologist sees this barren place as the jungle it once was.

The Jungles of Wyoming

Indeed, that is the startling conclusion from the geological evidence. Although Montana, North Dakota, and Wyoming are now semideserts, experiencing fierce blizzards in the winter, they were very different about 50 million years ago in the early Eocene. From plant fossils found in early Eocene rocks exposed along Puget Sound in Washington, to northern Oregon, the Bighorn Basin, and western North Dakota, we know these regions once resembled the rain forests of Central America or Vietnam. The most characteristic families of plants include the magnolias, the Rutaceae (the citrus family), the laurels (including relatives of avocado, sassafras, and the oriental camphor tree), the cashew family (including pistachio and mango trees), the tropical *Annona* trees (bearing fruit familiar in Central America as paw paws, cherimoya, custard apples, soursop, and sugarapple), the tropical vines of the moonseed family, and the icacina vines now found in subtropical Mexico (figure 1.2). Anyone who has ever tried to cultivate citrus, avocados, or mangoes knows that these plants need a lot of water and are especially sensitive to freezing.

Even extinct plants of the early Eocene with no living descendants can indicate climate. Jack Wolfe developed a

2

FIGURE 1.2. Living examples of some of the tropical plants which were fossilized in the lower Eocene beds of North America. (A) The flowering magnolia tree, typical of warm climates. (B) Branches and fruits of *Annona*, a genus that includes such tropical plants as the pawpaw. (Photo B courtesy B. Tiffney.)

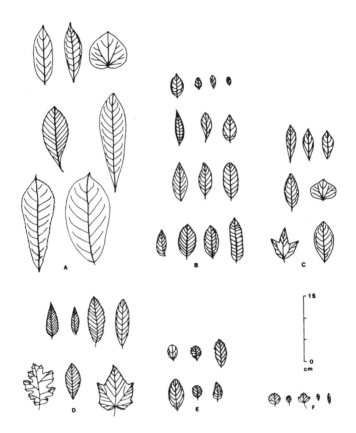

FIGURE 1.3. Spectrum of leaf shapes characteristic of different types of climate. (A) Tropical rain forest. Notice the large size, the smooth (entire) margin, the broad heart-shaped base and palmate veins; the drip tips are especially characteristic of rain forests. (B) Tropical semideciduous forest. Note the much smaller size and fewer drip tips. (C) Broad-leaved evergreen forest. (D) Low-temperature broad-leaved deciduous forest. Note the the many jagged margins, characteristic of cooler temperatures. (E) Dry temperate scrub. (F) Arctic vegetation, with the smallest leaves of all. (From Wolfe 1985; by permission of the American Geophysical Union.)

method for extracting climatic information from the size and shape of leaves, no matter what family of plants they belong to (Wolfe 1978, 1990). Tropical plants of modern rainforests have leaves of distinctive shapes. They tend to be larger, with smooth margins. Some are thick, indicating that they are not shed in the fall, but are evergreen. Many of the leaves found on vines and lianas have a heart-shaped base. Lower-story plants found in rainforests typically have "drip-tips" at the end of the leaf, so that abundant rainfall is shed easily. By contrast, plants from cooler climates have smaller leaves with jagged margins, and deciduous trees have thin leaves that grow quickly and can be shed economically (figure 1.3).

From these criteria, Wolfe (1978) estimates that mean annual temperatures in the early Eocene in the Pacific Northwest were 20-25°C (68–77°F), typical of the Central American tropics. More importantly, these regions had very small fluctuations in temperature, like the tropics. Wolfe estimates that temperature varied by only 5–10°C (7–15°F) over the year. The temperature of the coldest month seldom dropped below 18°C (64°F).

Leo Hickey (1977) analyzed the fossil flora of the early Eocene Golden Valley Formation of western North Dakota. He obtained a mean annual temperature estimate of around 18°C (64°F). The mean temperature in the coldest month was probably as high as 13°C (55°F) because many of the plants were intolerant of freezing. The fossil plant record also suggests that mean annual rainfall exceeded 150 cm (60 inches), so the region was well watered. By contrast, western North Dakota today has a steppe climate—the mean annual temperature is about 5°C (41°F), and the spread between daily extremes ranges over 33°C (over 90°F). In North Dakota or eastern Montana, it is not at all unusual for the temperature on a spring or fall morning to start out above 32°C (90°F) and then drop below freezing in a matter of hours as an Arctic cold front moves through.

From the paleobotanical evidence, we can visualize the Eocene Bighorn Basin as a tropical forest much like Panama. Tall trees formed a dense canopy, with vines and lianas growing in and around them. The red, gray, and greenish color bands that stripe the badlands slopes are more than just pretty

5

scenery. Each band represents an ancient soil horizon, and in some places there are hundreds of them stacked on top of each other spanning millions of years of the early Eocene. Each represents a new episode of floodplains awash in mud followed by soil growth and then flooding again. According to Tom Bown and Mary Kraus (1981, 1987), these ancient soils were deposited on broad floodplains bordering a meandering river system much like the tropical rivers of the Amazon basin. In the Powder River Basin of northeastern Wyoming, huge swampy deltas accumulated so much vegetation that they left coal seams over 100 meters (300 feet) thick! Although these coal seams are slightly older (Paleocene) than the early Eocene rocks of the Bighorn Basin, they are typical of the rocks that were formed in the ten million years after the extinction of the dinosaurs.

An Early Eocene Menagerie

Living in this tropical rainforest was an assemblage of animals as strange as the notion of a Wyoming rainforest (figure 1.4). Crocodiles and pond turtles were the most common reptiles, and frogs and salamanders thrived (Bartels 1980). The most surprising animal of all, however, was the giant "terror crane" *Diatryma*, a flightless predatory bird over 2 meters (7 feet) tall (figure 1.5). Its sharp beak was over 23 cm (9 inches) long, and it had a sturdy muscular neck for ripping flesh from its prey. It caught and held its prey with strong, sharp-clawed legs that were over 1.3 meters (4 feet) long. Such large predatory birds evolved several times in the geological past when there was no large mammalian predator to compete with them.

Indeed, the mammals of the early Eocene were very different from anything living today. Although there were no lion-size predators, meat-eating mammals known as creodonts (figure 1.4) were as big as modern wolves. The earliest members of the living order Carnivora were also present, although they were no bigger than a housecat. Although creodonts and carnivorans were both flesh eaters, they evolved most of their features independently. Both groups have slicing and stabbing teeth, but the details are different. For example, in the extinct order Creodonta, the enlarged shearing blades (known as carnassials)

FIGURE 1.4. Typical landscape during the early Eocene in North America. The jungle vegetation includes many ferns, palms, and magnolias. The snarling beasts in the front center are pantodonts of the genus *Coryphodon*, the largest mammal of the early Eocene. The predatory creodont *Oxyaena* is about to pounce from a rock on the left. A small, tarsier-like primate hides in the bush in the left foreground. Running in the background are two archaic hoofed mammals of the genus *Phenacodus*. (See also figure 1.6B). In the lower right is the earliest horse, *Protorohippus*. (Painting by R. Zallinger, courtesy Yale Peabody Museum.)

7

FIGURE 1.5. The giant eight-foot "terror crane" *Diatryma*, the largest predator during the early and middle Eocene in North America. (Painting by Z. Burian.)

C

D

FIGURE 1.6. (A) The dachshund-like archaic hoofed mammal *Hyopsodus*, common during the early and middle Eocene in North America. (From Gazin 1955.) (B) The sheep-size archaic hoofed mammal *Phenacodus*, one of the larger mammals of the early Eocene. (From Scott 1913.) (C) The earliest horse, *Protorohippus* (formerly known as *Hyracotherium* or "eohippus") from the early Eocene of North America. (Painting by Z. Burian.) (D) The earliest even-toed hoofed mammal, or artiodactyl, the rabbit-size *Diacodexis*. (Courtesy K. Rose.)

9

used for slicing flesh occur between the first upper and second lower molars, or the second upper and third lower molars. In carnivorans (members of the order Carnivora, including modern dogs, cats, bears, weasels, raccoons, mongooses, hyena, plus seals, sea lions, and walruses), the carnassial shear is developed between the last upper premolar and the first lower molar. Note that *carnivore* can refer to any flesh eater (including ourselves), but *carnivoran* is a member of the taxonomic order Carnivora.

Creodonts and carnivorans hunted a variety of mammals that are unrecognizable to us today. The dominant large herbivores were archaic hoofed mammals, only distantly related to living hoofed mammals such as horses or cows. The most common of these were the hyopsodonts (figure 1.6A), which had long bodies like dachshunds, and the phenacodonts, including the sheep-size beasts *Ectocion* and *Phenacodus*, the most abundant large mammals of the Bighorn Basin Eocene (figure 1.6B).

Phenacodonts were long thought to be ancestral to the perissodactyls, the order that includes horses, tapirs, rhinos and their extinct relatives. Often called the "odd-toed" hoofed mammals, most perissodactyls have only one or three toes on their feet. In fact, the early Eocene marks the immigration of the first true horses (figure 1.6C) into North America from China. Once known as *Eohippus*, these horses were the size of a cat, with four toes on the front foot and three toes on the hind foot. Although most textbooks now call this animal *Hyracotherium*, recent research by Jeremy Hooker indicates that *Hyracotherium* was an exclusively European mammal, and not even a horse (Hooker 1989); the proper name for the earliest American horse may turn out to be *Protorohippus*. Living side by side with the earliest horses was the earliest recognizable ancestor of tapirs and rhinos, a little beast known as *Homogalax*. This animal was almost indistinguishable from the earliest horse, except that its teeth had slightly stronger crosscrests for slicing leafy vegetation (in the manner of modern tapirs).

In addition to archaic hoofed mammals and early perissodactyls, there was a third group of hoofed mammals: the "even-toed" artiodactyls. Today, artiodactyls are represented by pigs, hippos, camels, antelopes, deer, giraffes, sheep, goats, and cattle. All of these animals have either two or four toes. In the early Eocene, we see the earliest representative of the artiodactyls, a jackrabbit-size animal called *Diacodexis* (figure 1.6D). It was like a jackrabbit not only in size, but also because it had long legs suited for jumping, similar to tiny antelopes such as the African dik dik. *Diacodexis*, primitive horses and tapirs, and phenacodonts make up most of the ground-dwelling herbivores, along with a few other archaic hoofed mammals that belonged to long-extinct groups unfamiliar to most of us.

Then there are the zoological puzzles. The largest mammal of the early Eocene was a cow-sized beast known as *Coryphodon*, last remnant of an extinct order of mammals known as pantodonts (figure 1.4). *Coryphodon* had a huge head with a broad snout and sharp canine teeth. Its molar teeth had sharp V-shaped crests like a giant insectivore, but it probably dined on leaves. It had broad hooves and a heavy build, yet there is no evidence it was related to true hoofed mammals (either living or extinct). Some have suggested that pantodonts were insecti-vores grown huge. Whatever pantodonts were related to, the early and middle Eocene was the end of their fifteen-million-year reign. *Coryphodon* bones and teeth are so large that even fragmentary pieces can be recognized while fossil hunting, simply by their size. Weathered piles of broken *Coryphodon* bones are so distinctive that paleontologists nicknamed them "*Corypho*-dumps."

In addition to the puzzling pantodonts, there were bizarre beasts known as tillodonts and taeniodonts (figure 1.7). Known mostly from skulls and teeth, tillodonts and taeniodonts are a complete zoological puzzle. Although they were big and herbivorous, they were probably not related to hoofed mammals; some had claws. The middle Eocene tillodont *Trogosus* and the taeniodont *Stylinodon* were almost bear-size, with chisel-like incisor teeth resembling those of a rodent. No one knows what these animals did for a living, nor are any bear-size gnawers living today to provide a clue.

In summary, the undergrowth beneath the early Eocene rainforest was populated by a strange mix of large herbivores.

11

A   B

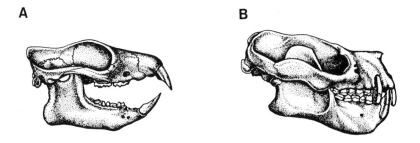

FIGURE 1.7. Tillodonts and taeniodonts, bizarre extinct mammals from the early Cenozoic whose relationships are unknown. (A) *Trogosus*, a bear-size tillodont from the middle Eocene, with huge gnawing incisors. The skull was over a foot long. (B) *Psittacotherium*, a wolf-size taeniodont from the Paleocene. Note the large chisel-shaped incisors and tusk-like canines. The skull was almost 10 inches in length. Both had heavy bodies and thick, clawed limbs; they may have resembled a cross between a bear and an aardvark.

A   B

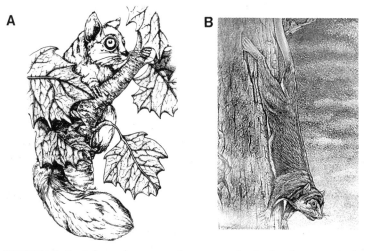

FIGURE 1.8. Arboreal mammals are particularly characteristic of the early Eocene jungles. (A) The tarsier-like primate *Tetonius*, one of the many archaic primates from the Bighorn Basin. (Courtesy K. Rose.) (B) The squirrel-like archaic group of mammals known as multituber-culates. Although very rodent-like in body shape, multituberculates originated in the Jurassic, and their closest relative may be the egg-laying platypus. (By permission of the American Association for the Advancement of Science.)

Extinct groups with no living relatives, such as pantodonts and tillodonts, browsed alongside extinct archaic hoofed mammals like hyopsodonts and phenacodonts. There were also the most primitive members of the living orders of hoofed mammals, the artiodactyls and the perissodactyls. Almost all of these beasts had simple, low-crowned molar teeth for browsing on the abundant leafy vegetation.

Yet the most numerous herbivorous mammals were not on the ground. The rainforest canopy was an ideal habitat for tree dwellers, and there mammals lived in abundance. When we think of modern rainforests, we think of monkeys, and indeed the Eocene rainforests had primates. Distantly related to the living lemurs of Madagascar, these primitive Eocene primates had the well-developed eyes, opposable thumbs for grasping, and long tails typical of primates (figure 1.8A). Yet they lacked many of the advanced specializations of monkeys and apes; in particular, they did not have unusually large brains. Primates were a great success in North America in the Eocene, but today North America has no native monkeys or lemurs. The human primate indeed dominates the continent, but we are not natives.

Next in importance to primates was a group of mammals that arose early in the age of dinosaurs. Known as multituberculates (figure 1.8B), they had squirrel-like bodies with long prehensile tails. Their skeletons show that they were very archaic mammals, distantly related to the living platypus, which lays eggs. Multituberculates get their name from their molar teeth, which are unusually long grinding batteries with multiple bumps, or "tubercles." Characteristic are the blade-shaped premolars that sat at the front of their lower grinding teeth; these may have been used to slice through the hulls of tough nuts and seeds. Multituberculates also had chisel-like incisors, just like rodents, which helped them gnaw seeds and fruits as squirrels do today. Clearly, they filled the ecological niche of rodents through almost 100 million years, from early in the age of dinosaurs through the Paleocene.

In the early Eocene, however, multituberculates were exposed to competition with the first true rodents. Migrating from Asia just before the Eocene, rodents soon became common in

13

North America, and quickly began displacing the archaic multituberculates and the squirrel-like primates. The earliest North American rodents were very primitive in most features, although they did have chisel-like incisors. Squirrels might be the closest living analogue.

In short, an early Eocene menagerie would not be at all familiar to the average zoo visitor. Most of the animals were members of extinct orders and families with no living descendants. Those which have living relatives (the earliest horses, tapirs, artiodactyls, and carnivores) would scarcely be connected with anything in the zoo today. Only the squirrel-like rodents and the lemur-like primates would bear faint resemblance to their living descendants.

Clearly, these animals were mostly leaf- and fruit-eaters adapted to living in dense jungles. The abundance of tree dwellers shows that there must have been a tall canopy of trees, typical of moist tropics. The presence of crocodilians, pond turtles, frogs, and salamanders also suggests that the region was well-watered with wide floodplains and swampy areas.

These qualitative aspects of the fauna can be complemented by quantitative techniques. For example, plotting the body weights of all the animals in sequence produces a cenogram (figure 1.9A). The shapes of cenograms are characteristic of different types of habitats. Forest and jungle faunas have many different animals in the middle range of body sizes, and only a few large animals (Legendre 1987, 1988). The cenogram of a modern jungle fauna thus has a very flat slope, with only a short peak for the largest animals. By contrast, cenograms for savannahs and open woodlands have more large-bodied animals, but fewer total species, especially in the middle size ranges. Consequently, such cenograms have a sharp break between the small and large body size lines. A cenogram for the early Eocene mammals of the Bighorn Basin (Gingerich 1989) clearly indicates that they lived in tropical, humid jungles (figure 1.9B).

Another way of analyzing the faunas is to break them into dietary and locomotory categories, following a system developed by Peter Andrews (Andrews et al. 1973). Jeremy Hooker has used this method to analyze the mammals of the early Eocene London Clay (Collinson and Hooker 1987). According

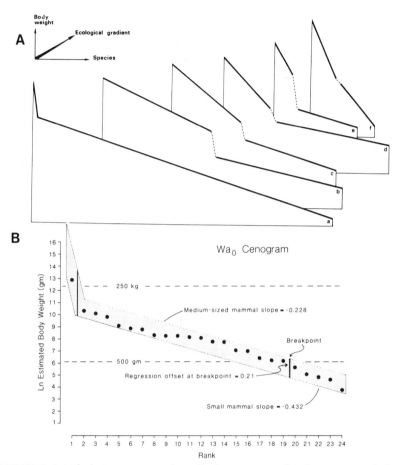

FIGURE 1.9. (A) Cenograms plot all the species of a fauna in rank by body weight, with the heaviest on the left. The longer the line, the greater the diversity. The steeper the slope, the fewer mammals there are in each size class. A gap in the medium-size species indicates an open environment, such as a mixed savanna-woodland. In the foreground is a plot of a tropical rainforest environment (in Gabon), followed by a tropical woodland savanna (in Rwanda), and tropical savanna (in Zaire), a semidesert (in Algeria), a desert (in Iran), and a Mediterranean scrub (in Spain) in the background (Modified from Legendre 1986). (B) Cenogram of earliest Eocene mammals from the Bighorn Basin. Note that the shape most closely resembles the tropical rainforest of Gabon. (From Gingerich 1990; by permission of the University of Michigan Museum of Paleontology).

to Andrews's categories, the dominant types were arboreal mammals (mostly primates and multituberculates), insectivores, small ground mammals (mostly rodents and smaller archaic hoofed mammals), and browsing herbivores (especially pantodonts, primitive perissodactyls, artiodactyls, and larger archaic hoofed mammals). *Hyracotherium* (the name erroneously applied to early Eocene horses) was among the most common mammals. Most of the primates and rodents were fruit eaters. According to Andrews's methods, the distribution of these feeding and locomotory types most closely matches faunas found in lowland evergreen forest, such as is found today in Borneo or central Africa.

The paleobotanical evidence from the London Clay supports this interpretation. As described by Margaret Collinson, early Eocene floras of the London Clay consist of tropical trees, shrubs, and lianas, many of which were members of the families that today include such trees and shrubs as cinnamon, figs, magnolias, palms, laurels, citrus, paw-paw, incense and tea and cashew families, and vines such as moonseed, icacina, and grape. Collinson showed that 92% of these plants have nearest living relatives that are tropical, mostly in the rain forests of southeast Asia (Collinson 1983; Collinson and Hooker 1987). Fringing the coasts of this tropical forest were mangrove swamps full of the *Nypa* palm (figure 1.10A). Altogether, the London Clay produces about 350 species of plants, most of which have nearest relatives in the Malay Peninsula. This fact joined with other lines of evidence indicates that the average Eocene temperature in what is now London was about 25°C (77°F), compared to the modern average of 10°C (50°F). Instead of the cold, foggy, clammy London of the Sherlock Holmes stories, Eocene London was as warm and tropical as Singapore!

The Palmy, Balmy Polar Regions

If tropical forests in Montana or London are not startling enough, even more astonishing are the early Eocene fossils from Alaska, Arctic Canada, Siberia, and Spitsbergen. Even though most of these localities are near the Arctic Circle, their Eocene fossils indicate unusually warm conditions. At latitudes

A

B

FIGURE 1.10. Tropical plants found in many lower Eocene beds in high latitudes. (A) Swamp in Florida with *Nypa* palms. (B) The archaic gymnosperms known as cycads, or "sego palms." (Both photos courtesy B. Tiffney.)

above 61° north, floras were dominated by broadleaved evergreens, especially palm trees and cycads, the palm-like gymnosperms that were so common in the age of dinosaurs (figure 1.10B). According to Jack Wolfe (1980), these floras indicate a mean annual temperature that is downright balmy: 18°C (65°F).

North of these latitudes were broad-leaved deciduous forests typical of temperate regions, now found in an area that experiences six months of darkness. Fossil floras show that in the Eocene Spitsbergen could not have suffered freezing temperatures (Schweitzer 1980). Ellesmere Island, the northernmost point in the Canadian Arctic, is now at a latitude of 78° north, well within the Arctic Circle and only a few hundred miles from the North Pole. This locality yields abundant fossil plants—even coal beds. In addition to fossil plants, there are fossils of alligators, pond turtles, land tortoises, and monitor lizards, as well as garfish and bowfins (Estes and Hutchison 1980). Most of these animals are typical of the warm climate of the southeastern United States. Monitor lizards are found in the tropics of southeast Asia. None can tolerate freezing temperatures for very long. Alligators are limited by mean coldest month temperatures of 10°C (50°F).

The Eocene mammals of Ellesmere Island were also typical of the tropical early Eocene assemblages, including primates, rodents, creodonts, *Coryphodon*, primitive perissodactyls, and multituberculates (McKenna 1980). Surprisingly, the most common animals were extinct relatives of the living colugo, a gliding mammal distantly related to bats that is erroneously called the "flying lemur." Colugos today are found only in the tropics of southeast Asia, particularly Indonesia, Malaysia, and the Philippines. Clearly, the Canadian Arctic was warm and temperate during the early Eocene, supporting subtropical-temperate animals and plants, and it never experienced significant freezing.

How could this be? If the area was above the Arctic Circle, then it must have had six months of darkness. Analysis of the paleomagnetic pole positions for Eocene rocks of the North American continent show that these regions were not significantly further south in the Eocene, so we can rule out subsequent northward drift of the continent from a temperate latitude

(McKenna 1980). According to McKenna's (1983b) calculations, Ellesmere Island was at 75° north latitude, not significantly different from today. Wolfe (1980) and others have suggested that the Earth's axial tilt might have been less than present, making the Arctic Circle smaller and exposing even 78° north latitude to year-round sunlight. However, geophysicists have argued that such a radical change in the earth's tilt is difficult to explain given all we know about physics and planetary motions (Ward 1982; Harris and Ward 1982). So drastic an alteration of the earth's angular momentum would have disrupted life far more profoundly than is borne out by the geological record. An examination of the geological record, including the seasonality shown by Precambrian cyanobacterial mats (Vanyo and Aramwik 1982) indicates that the earth's spin axis has remained remarkably constant. Barron (1984) modeled the climate of an earth with less axial tilt and produced polar cooling, not polar warming. Creber and Chaloner (1984) examined the botanical evidence, and concluded that even with six months of darkness, the polar regions received sufficient sunlight for a seasonally productive forest; temperature would have been the limiting factor. The dominance of tundra vegetation in these regions today is dictated by temperature, not darkness. In addition, fossil wood from Ellesmere Island, Greenland, and Spitsbergen shows pronounced growth rings, indicating a strongly seasonal climate. This would not be so if the Earth's tilt had decreased.

McKenna suggested that the amount of radiation from the sun might have been greater during the Eocene (McKenna 1980). Unfortunately, there is no way of testing this hypothesis. Whatever the explanation, most scientists have become reconciled to the fact of global Eocene climates so warm that summer heat stored in oceanic waters kept the Arctic region above freezing even in winter. The region would still have undergone six months of darkness, but most Eocene reptiles from Ellesmere Island were capable of hibernating, and many of the plants could have gone dormant in the dark, as long as they did not freeze.

If the Arctic was balmy in the Eocene, what about the Antarctic and high latitudes of the southern hemisphere? Elizabeth Kemp (1978) has shown that the early Eocene floras of Australia were mostly tropical, too. Relatives of many of the plants are found today only in the jungles of New Guinea and New Caledonia. Other fossils are typical southern hemisphere plants (figure 1.11), such as *Araucaria* (the Norfolk Island pine and the "monkey puzzle tree") and the typical southern conifers, the podocarps. The family Proteaceae (including such colorful flowering shrubs as *Banksia*) was abundant, as were *Nypa* palms (figure 1.10A). In some places, the conditions were so swampy that coal beds formed. The great diversity of fungi and fern spores shows that conditions were very wet; some estimates of rainfall exceed 150 cm (60 inches) a year. All of the evidence suggests a dense rainforest canopy watered by warm, moist winds with a sluggish, irregular circulation pattern over Australia. The offshore waters averaged 20°C (68°F)—much warmer than the present waters off Tasmania and southern Australia.

Even Antarctica was unusually temperate. There is no evidence of significant glaciers in the early Eocene. The plants were typical cool-temperate southern hemisphere varieties, including the southern beech *Nothofagus* (figure 1.11C), podocarps, araucarias, and abundant ferns (Case 1988). Eocene fossils of *Nothofagus* are particularly good indicators of climate, since most members of this genus today live in cool temperate rain forests in South America, New Zealand, Australia and Tasmania. Although fossil collecting is difficult in Antarctic conditions, there have been some successes. The marine molluscs recovered from Seymour Island in the Antarctic Peninsula indicate that the offshore waters were a cool-temperate "Weddell Province," distinct from the warmer waters off Australia (Zinsmeister 1979, 1982). In 1981 the first fossil mammal was recovered from the Eocene beds of Antarctica (Woodburne and Zinsmeister 1982; Woodburne 1984). It was a member of a peculiar extinct group of pouched mammals, or marsupials, that are known from South America in the Eocene. Because Antarctica, Australia, and South America were once all part of the great Gondwana continent, and were still connected in the early Eocene, these pouched mammals must have inhabited the cool-temperate *Nothofagus* forest from

20

FIGURE 1.11. Typical plants of the southern hemisphere temperate zone. (A) The primitive conifer *Araucaria*, familiar as the Norfolk Island pine, or monkey-puzzle tree. (Photo courtesy C.R. Prothero). (B) Members of the Australian family Proteacea include *Banksia*. (C) The "southern beech," *Nothofagus*.

southern South America to Antarctica. Recently, early Eocene mammals were described from Australia. They included not only the expected primitive marsupials, but also a very primitive placental mammal that suggests even more of South America's fauna made the trip over Antarctica to Australia (Godthelp et al. 1992).

In summary, both polar regions were cool and temperate during the early Eocene, with no hint of the great ice caps soon to come. Cool-temperate forests flourished at both poles, even though much of the region experienced six months of darkness. In the Arctic, alligators and pond turtles thrived, along with a variety of typical early Eocene mammals. During the dark months, they must have hibernated or migrated south. Fossils from the Antarctic are scarce, but apparently there was a cool temperate plant community supporting the pouched mammals from South America. Clearly, the globe was much warmer in the polar regions than at any time since then. But what about the tropics?

The Tropical Tethys
In the early Eocene, a great equatorial seaway sprawled eastward from the Mediterranean to southeast Asia. Known as Tethys, this great warm-water belt was a dominant feature in Eurasia for over 300 million years (figure 1.12). Because Africa and Arabia had not yet collided with Europe, nor India with Asia, there was an unbroken marine connection from the Atlantic to Indonesia. Warm, shallow waters circulated in the Mediterranean and across the Indo-Pacific region. High sea levels during most of the early and middle Eocene inundated large areas of Europe, northern Africa, Arabia, and the middle East with shallow epicontinental seaways, in which huge quantities of shallow marine limestone accumulated. A number of lines of evidence suggest that these seaways were tropical, with average water temperatures over 25°C (77°F), much like the modern tropical belt. However, as we have already seen, tropical conditions extended well north and south of the modern equatorial region, so that warm-water organisms were found as far as 50° south of the Eocene equator in places like western Australia, and 50°N of the Eocene equator in northern Europe

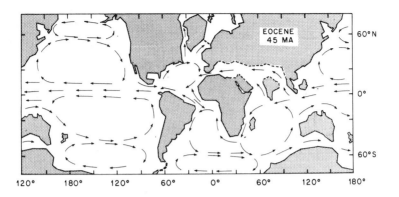

FIGURE 1.12. Distribution of the continents and patterns of inferred oceanic circulation during the middle Eocene, about 45 Ma. Note the great tropical Tethyan seaway running from the Mediterranean through the gap north of India to Indonesia. There was also no circum-Antarctic circulation. Instead, polar waters flowed around the southern oceans, and exchanged heat with more temperate and tropical waters. (Modified from Frakes 1979.)

(Adams et al. 1990). Today, these regions are in the cold temperate belt.

The most abundant invertebrates in these Tethyan seaways were molluscs similar to those found in our shallow oceans today. A great diversity of fossil clams and snails occur in early and middle Eocene deposits. Most of them are closely related to living species that prefer warm-water conditions. After molluscs, the next most common marine invertebrates were the echinoids (sea urchins and their relatives). Early and middle Eocene echinoids were richly diverse, and again they were mostly warm-water species. In the early Eocene the first sand dollars, their flattened forms adapted to burrowing rapidly in shifting nearshore sandy deposits, evolved—probably in Africa or India—from more spherical ancestors. By the middle Eocene, sand dollars had spread worldwide.

Coral reefs, which had been severely affected by extinctions at the end of Age of Dinosaurs, began to recover in the Paleocene. By the early Eocene, widespread coral reefs abounded

23

all along the Tethyan belt, and solitary corals were common in regions outside the tropics. Modern reef-building corals survive only in warm oceans. Symbiotic algae living in their tissues provide oxygen, and also make it easier for corals to secrete their limestone skeletons. Because these algae are plants, however, the symbiotic corals cannot live in waters that are too deep or muddy to provide enough light. Hence, corals are restricted to warm (never colder than 16°C, or 61°F), shallow waters without too much mud to clog them. It is not surprising then to see reef corals so abundant in the Tethys, but to find them 10° of latitude further north and south than today shows that even middle latitudes were nearly tropical (Adams et al. 1990).

In addition to familiar animals like clams, snails, corals, and sea urchins, Eocene oceans hosted some unfamiliar organisms as well. Amoeba-like protozoans known as foraminiferans are among the most common single-celled fossils preserved in marine sediments. There are two basic ecologies adopted by foraminiferans. Some are so tiny that they can float freely in the open ocean, capturing smaller microorganisms on which they feed. These are known as planktonic foraminiferans, and they have been one of the most common and important microfossil groups in the oceans since they arose in the Cretaceous. Others live on the ocean bottom and creep along, capturing food particles with long finger-like projections of protoplasm. These bottom dwellers, or benthic foraminiferans, secrete a shell made of calcite that is rarely more than a few millimeters long.

In the early and middle Eocene, however, one group of benthic foraminiferans created shells of exceptional size. Now extinct, these nummulitids (figure 1.13A) secreted a disc-shaped shell made of tiny spiraling chambers. The amoeba-like animal presumably lived only in the last chamber. Fossils of nummulitid discs sometimes reach several centimeters in diameter—an immense skeleton for a single-celled organism. Paleontologists have long puzzled over why such a tiny organism secreted such a large shell. The most likely explanation is that nummulitids, like many large benthic foraminiferans alive today, probably incorporated symbiotic algae in their tissues to help them secrete calcite.

FIGURE 1.13. The coin-sized foraminiferans known as nummulitids were common in the tropical Tethys during the Eocene. These shells were among the largest fossils known from a single-celled, amoeba-like group of protozoans. (A) Close-up of a sectioned specimen showing the chambers arranged in a flat spiral. (B) Some Eocene limestones are solid nummulitid discs. (Photo courtesy T. Aigner.) (C) Nummulitic lime stones were especially common in the Mediterranean region. The Pyramids were built mostly of nummulitic limestones from the quarries (foreground) of the Gizeh Plateau near Cairo, Egypt.(Photo courtesy R. Schoch.)

25

Thus, it is not surprising that nummulitids, like corals, preferred the tropical Tethyan belt. However, their abundance in the shallow Tethyan seaways is truly staggering. Many limestones in the region are made almost exclusively of nummulitid discs; some limestones resemble a bunch of coins glued together (figure 1.13B). Indeed, the Latin word "nummus," is a kind of ancient coin. In some older texts, stratigraphers call the Eocene the "Nummulitic Period." Thick nummulitic limestones are among the few suitable building materials in Egypt; the great pyramids are built partly of the skeletons of nummulitids (figure 1.13C). When the Greek historian Herodotus visited the pyramids, he noticed the abundant discs all over the ground—and thought they were petrified lentils from the lunches of the slaves who built the pyramids. In the Paris basin, the coin-like nummulitids gave some middle Eocene limestones the name "banc royal."

Protozoans are near the base of the food pyramid. What about animals at the top? A great variety of bony fish fossils are known from Eocene deposits such as Monte Bolca in Italy. Although most of the fish represent extinct species, they were morphologically similar to many of the fish now found in the coral reefs and tropical waters. Clearly, the warm tropical waters of Tethys were a haven for fish, just as we might find in the Caribbean or the Great Barrier Reef of Australia today. In addition to bony fish, there were sharks and rays very similar to living species.

When the last of the giant marine reptiles had vanished at the close of the Cretaceous, they left a void. Except for sharks, there were no large marine vertebrate predators in Paleocene seas. But the Eocene marks the origin of the largest living animals, the whales. Eocene whales, known as archaeocetes, were very different from their modern descendants (figure 1.14). Although archaeocetes were large and aquatic, they did not yet have all the specializations we associate with modern whales. Some reached a length of 80 feet (24 m), and weighed about 12,000 pounds (5400 kg). The archaeocete *Basilosaurus* reached 14 meters (about 47 feet) in length, and had well-developed flippers as its front limbs. Specimens recently discovered in Egypt show that it still had tiny vestigial hind limbs.

26

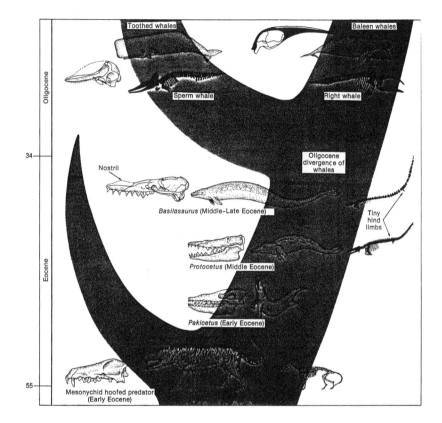

FIGURE 1.14. Evolution of whales from mesonychids. Early Eocene beasts, such as *Pakicetus*, had teeth and ears like mesonychids, but a whale-like skull. Middle Eocene whales such as *Basilosaurus* were aquatic, but still had tiny, vestigial h ind limbs; other middle Eocene whales may have had flippers on their hands but retained walking hind limbs. By the early Oligocene, the archaic archaeocete whales had been replaced by their two living descendants, the toothed whales and the baleen whales.

27

According to Philip Gingerich and others who described it, these hind limbs were too small to carry any weight. These authors suggest that the hind limbs may have been retained to help the males clasp females when mating in water (Gingerich et al. 1990). Archaeocete skulls were equipped with a long fish-catching snout, but they had no blowhole on the top of the head; the nostrils were at the end of the snout. Archaeocetes did not have the specialized whale ear, so they could not navigate by echo sounding, as modern whales do. Most modern toothed whales have hundreds of peg-shaped teeth in their mouths for gripping fish, but archaeocetes had broad, bladed teeth with serrated edges.

So where did whales come from? Those odd teeth give the clue to their ancestry. Broad triangular teeth with serrated edges most closely resemble those of an extinct group of Paleocene and Eocene hoofed mammals called mesonychids (figure 1.14). Most mesonychids are reconstructed as wolf-like or bear-like, but the detailed anatomy of their skeletons reveals that they were carnivorous or scavenging hoofed mammals. Beyond the teeth, there are other similarities in the skull and ear regions of whales that show they descended from terrestrial mesonychids.

An even more primitive whale was discovered by Philip Gingerich, Donald Russell, and colleagues in 1983 (Gingerich et al. 1983). Known as *Pakicetus*, it comes from early Eocene deposits in Pakistan that were right on the margin of the Tethys seaway. *Pakicetus* is known mostly from its skull, and that skull has even more mesonychid-like teeth and a primitive ear region incapable of echo-sounding. Gingerich and colleagues speculate that *Pakicetus* lived by feeding on fish on the shores of Tethys. However, without a skeleton, we cannot determine whether it had hooves or flippers. Recently Gingerich and his colleagues have reported a middle Eocene fossil whale from west Africa which apparently had flippers for its front limbs, but fully terrestrial hind feet.

The notion of a terrestrial animal becoming aquatic is actually not so far-fetched, since it has happened numerous times. After all, seals evolved from the bear family in the early Miocene. The great marine reptiles—ichthyosaurs, plesiosaurs, mosasaurs—of the Age of Dinosaurs were also descended from terrestrial reptiles. Earl Manning suggested that the living brown hyaena might be a good analogue for the ancestor of whales. Although it is a terrestrial carnivore with paws, it is known as the "strand wolf" in South Africa for its habit of roaming seashores in search of fish and carrion. A mesonychid with a similar habitat could have become progressively more aquatic as it spent more time in the water catching living fish, rather than waiting for dead things to wash ashore. Some modern bears are adept at swimming and catching fish, and this might explain why seals, sea lions and walruses are descended from a bear-like ancestor.

By the middle Eocene, archaeocete whales ranged all over the world, from Great Britain and North Africa, to the United States, Australia, New Zealand, Antarctica, and India. However, the fact that the most primitive ones are found in Egypt and Pakistan suggests that they originated on the shoreline of the ancient Tethys sea. Thus, the Tethys was the "Garden of Eden" for much of early Eocene life. It was an immense tropical belt that spanned three continents, and influenced conditions far to the north and south. Even regions far from Tethys were warmer than now, and were populated by warm-water organisms which had spread from a Tethyan source.

An Eocene Garden of Eden?
The temperate nature of the Arctic not only made it a haven for mammals, but also facilitated their transcontinental migration. McKenna (1979, 1983a, 1983b) has shown that there was still a connection between North America and Europe in the Paleocene and early Eocene along the Greenland-Scotland Ridge, allowing mammals to pass freely from one continent to another (figure 1.15). About half of the mammals found in the lower Eocene rocks of North America also appear in Europe, including multituberculates, creodonts, carnivorans, rodents, insectivores, primates, hyopsodonts and phenacodonts, tillodonts, *Coryphodon*, and even the artiodactyl *Diacodexis*. Although the earliest horses in North America have been called by the European name *Hyracotherium*, Hooker (1989) has shown that true *Hyracotherium* was not a horse, but a member of a closely

FIGURE 1.15. There were several possible land routes between the Holarctic continents in the early Eocene, including two potential North Atlantic corridors and the Bering land bridge. The great similarity between early Eocene faunas of North America and Europe suggests that the North Atlantic route was very effective. However, by the middle Eocene, this North Atlantic connection was severed, and Europe began to develop its own endemic faunas. Europe was also isolated from Asia by the Obik Sea, and from Africa by the Tethys sea. The Bering route was used throughout the Cenozoic, and there were many periods of faunal interchange between Asia and North America. (Modified from Savage and Russell 1983.)

related perissodactyl group, the palaeotheres. Nevertheless, the similarity between early Eocene European and American perissodactyls is striking enough that they were once placed in the same genus.

Indeed, the cosmopolitan nature of the early Eocene fauna was even more striking because artiodactyls, perissodactyls, tillodonts, advanced lemur-like primates (known as "euprimates," including the families Adapidae and Omomyidae), hyaenodont creodonts, opossums, and rodents all appeared in North America during the late Paleocene and early Eocene. For years, paleontologists have speculated on the origin of these orders. Some have pointed to Central America, or southeast Asia, or a number of other areas that have not yet produced many Paleocene mammals. Recently, however, strong evidence has emerged that many of these mammals arose in Asia or around the Tethys seaway. *Radinskya*, the oldest known relative of the perissodactyls, was recently described from the late Paleocene of China (McKenna et al. 1989). Together with *Minchenella*, the oldest known relative of the elephants and their kin, these discoveries suggest that the higher ungulate clade (elephants, plus sea cows, hyraxes, and perissodactyls) arose during the late Paleocene in Asia or the Tethyan region, where they became widespread by the Eocene (Prothero and Schoch 1989).

The most primitive known artiodactyl, *Diacodexis pakistanensis*, first appears in the early Eocene of Pakistan (Thewissen et al. 1983). Specimens discovered in the last few years show that both rodents and rabbits originated from a common ancestor which lived during the Paleocene in China (Li and Ting 1985; Novacek 1985; Luckett and Hartenberger 1985; Dashzeveg and Russell 1988). Tillodonts were also diverse in Paleocene deposits of China (Li and Ting 1983; Russell and Zhai 1987) and in the Eocene rocks of Pakistan (Lucas and Schoch 1981), and most authors now regard them as originating in Asia (Chow and Wang 1979; Gingerich and Gunnell 1979; Sloan 1987). Some authors (e.g., McKenna 1967; Gingerich 1986; Franzen 1987) argued that euprimates were African in origin, although Szalay and Li (1986) suggested an Asian origin. Possible omomyid primates have recently been discovered from the late Paleocene (Capetta et al. 1987; Gheerbrant 1987) and early Eocene (Hartenberger et al. 1985) of Africa, supporting the former hypothesis. Clearly, these common patterns suggest an east Asian or Tethyan source for all of these critical groups of mammals. They did not evolve from North American natives in the Paleocene, as originally proposed by Sloan (1969) and Gingerich (1976, 1977).

The suddenness of their immigration outside Asia has led some to argue that Eocene mammals may have been further isolated. India, which had been an island continent drifting away from its Gondwana source since the Cretaceous, collided with Asia about the beginning of the Eocene (figure 1.16). Some paleontologists (McKenna 1983b; Krause and Maas 1990) have suggested that India was analogous to "Noah's ark"; its own endemic mammals evolved in isolation during the Paleocene. When India docked with Asia in the Eocene, these advanced mammals spread out of Asia. The sudden appearance of the most primitive known artiodactyls in the early Eocene of Pakistan might support this idea, but primitive relatives of perissodactyls, proboscideans, tillodonts, euprimates, rodents, and rabbits were already in China in the Paleocene. Unfortunately, we do not have any Paleocene mammals from India to test the hypothesis. Eocene mammals from India look much like those of the rest of the Tethys (Russell and Zhai 1987).

During the Paleocene, eastern Asia was isolated from the rest of the world. It possessed a fauna composed mostly of endemic orders and families of mammals, including a great diversity of pantodonts, the puzzling rodent-like didymoconids, the bizarre multihorned uintatheres (discussed later), a tremendous diversity of groups related to rodents and rabbits, plus diverse mesonychids (Li and Ting 1983; Russell and Zhai 1987). There were a few types of multituberculates and archaic hoofed mammals in eastern Asia, but these could have been relicts from the Cretaceous or early Paleocene. There was a possible connection to South American mammals (Gingerich 1985), suggested by a strange group of South American native hoofed mammals, the arctostylopids, and a possible relative of the edentates (anteaters, sloths, and armadillos) known as *Ernanodon*. However, the late Paleocene of Asia had no genera in

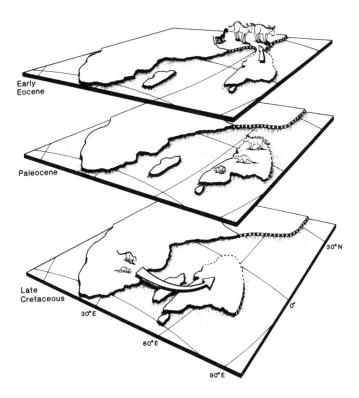

FIGURE 1.16. India split from the other Gondwana continents in the Cretaceous, and may have served as a "Noah's ark," docking with Asia in the early Eocene to release mammals that had evolved in isolation (including perissodactyls, artiodactyls, several groups of primates, and the hyaenodont creodonts). (From Krause and Maas 1990; by permission of the Geological Society of America.)

common with any other continent, and this endemism was almost as extreme in the early Eocene.

If a late Paleocene Asian origin is accepted for many Holarctic mammals of the Eocene, then we need to examine why these animals did not leave Asia until the early Eocene. During the Eocene, eastern Asia was connected to North America via the Bering land bridge, but McKenna (1983b) argues that the bridge was at too high a latitude to allow much dispersal. However, we have just seen that Ellesmere Island was probably at 75° ± 3° north paleolatitude in the Eocene, and it supported a significant mammal fauna. Parrish (1987) sees no reason why the Bering connection would have been any less hospitable for migration than was the North Atlantic corridor, and Hillhouse and Grommé (1982) calculated a paleolatitude of 83° ± 9° north for the late Paleocene Cantwell Formation of central Alaska. As we shall see shortly, the Bering corridor was indisputably effective in the middle and late Eocene, and its paleolatitude had not changed much since the early Eocene. If anything, it might have become less hospitable in the late Eocene because of climatic cooling.

Along the present Ural Mountains, the Obik Sea cut Asia in half (figure 1.15), connecting the Arctic Ocean with the Tethys, restricting terrestrial dispersal from east Asia to western Europe (although there was a connection with eastern Europe). However, there were several corridors across these seaways that might have allowed Eurasian mammals access to Europe, and then across Greenland to North America in the early Eocene (McKenna 1983b).

Africa was apparently isolated from much of the rest of the world during the Paleocene and Eocene, although its sparse mammal faunas of the late Paleocene (Capetta et al. 1978; Gheerbrant 1987) suggest that some kind of trans-Tethyan dispersal was occurring. By the early Eocene (Hartenberger et al. 1985) and Oligocene (Rasmussen et al. 1992), we find predominantly endemic mammals, especially primitive proboscideans and their kin, diverse hyraxes, elephant shrews, peculiar rodents, and some of the earliest monkeys. These mammals suggest that Africa became isolated by the Eocene after having received Tethyan immigrant ancestors for its endemics; the nearest relatives of proboscideans, arsinoitheres, hyraxes, elephant shrews, euprimates, and African rodents all occurred in China or along Tethys in the late Paleocene. Perhaps these immigrants from the north reached Africa by swimming and island hopping across Tethys in the late Paleocene (Gheerbrant 1987), then became more isolated and evolved into distinctive groups by the early Eocene.

No matter where we look, it is clear that the early and middle Eocene was a lush, tropical world, very different from today. Mammals dispersed across the North Atlantic, all along the Tethys from Asia to Africa and Europe, and possibly over the Bering Straits as well. In many ways, Eocene climate and vegetation was similar to the middle Cretaceous, when shallow seaways covered the continents, and warm climates prevailed. In both cases, scientists have suggested that such unusual warming was due to an excess of certain gases (especially CO_2) which generates a "greenhouse effect," letting in sunlight but trapping heat radiated back from earth. If this is so, then we might justifiably call the early Eocene a reprise of "the greenhouse of the dinosaurs." The final chapter will examine the various explanations for Eocene warming. The next chapter, however, will straighten out some confusion about time scales.

FIGURE 2.1. Charles Lyell as he appeared in 1836, shortly after publishing the first edition of *Principles of Geology*, which established the Cenozoic time scale.

Dawn of the Recent

The period next antecedent [to the Miocene] *we shall
call* Eocene, *from* ἡώs, aurora, *and* καιν όs, recent,
*because the extremely small proportion of living species
contained in these strata indicates what may be
considered the first commencement, or dawn, of the
existing state of animal creation.*

—CHARLES LYELL, *PRINCIPLES OF GEOLOGY* (1833)

Before proceeding further with the story of the Eocene greenhouse, I will take a moment to discuss the time scale itself. Much of the confusion about the Eocene-Oligocene transition is the result of arguments and misunderstanding about the time scale, which has been changed frequently in the last thirty years. Many of the hypotheses about asteroid impacts, periodic extinctions, or catastrophic events at the boundary were erroneous because they were based on an incorrect chronology. The time scale has changed so radically in just the last several years that most publications prior to 1990 are badly out of date. In many cases, what were once identified as late Eocene rocks are now middle Eocene. Some "early Oligocene" terrestrial rocks in publications of 1989 are now late Eocene, and most "middle Oligocene" terrestrial rocks are now early Oligocene.

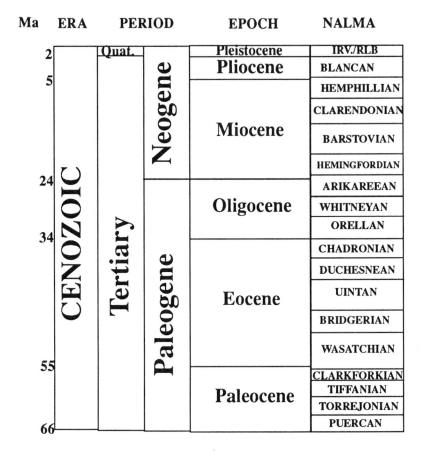

Ma	ERA	PERIOD		EPOCH	NALMA
2	CENOZOIC	Quat.		Pleistocene	IRV./RLB
		Tertiary	Neogene	Pliocene	BLANCAN
5				Miocene	HEMPHILLIAN
					CLARENDONIAN
					BARSTOVIAN
					HEMINGFORDIAN
24			Paleogene	Oligocene	ARIKAREEAN
					WHITNEYAN
34					ORELLAN
				Eocene	CHADRONIAN
					DUCHESNEAN
					UINTAN
					BRIDGERIAN
55					WASATCHIAN
				Paleocene	CLARKFORKIAN
					TIFFANIAN
					TORREJONIAN
66					PUERCAN

FIGURE 2.2. The modern Cenozoic time scale. Tertiary and Quaternary were the original period subdivisions, but some authors prefer the more symmetrical divisions of Paleogene and Neogene. The Eocene, Miocene, and Pliocene Epochs were proposed by Lyell (1833); the rest came later (see text). The modern correlation of the North American land mammal "ages" (NALMA) is also shown. Abbreviations: Ma = million years before present; Quat. = Quaternary Period; IRV./RLB. = Irvingtonian and Rancholabrean land mammal "ages" of the Pleistocene.

The rocks we now call Tertiary (figure 2.2) were one of the last parts of the geological time scale to be subdivided (Berry 1987; Schoch 1989b; Prothero 1990). In the eighteenth century, European geologists and miners recognized three basic divisions of the rock column. In the cores of mountain ranges, the igneous and metamorphic rocks that formed the "basement" were called "Primary" or "Primitive." These rocks were thought to have formed chemically during the disorder of the first days of creation. "Primitive" rocks were overlain by fossiliferous sedimentary rocks, which were often steeply dipping or highly deformed. Called "Secondary," these strata were thought to have been deposited during Noah's flood. In the low hills surrounding these mountains were loosely consolidated but stratified clays, sands, and limestones, full of fossils that resembled living species. For this reason, these "Tertiary" sediments were thought to be post-Flood deposits. This simple, three-fold division of geologic time (formalized in northern Italy by Giovanni Arduino in 1759) worked in a very crude sense during the eighteenth century.

As the geologic time scale was further refined, such simplistic divisions began to break down. Eventually, the concept of "Primitive" rocks was superseded by the formal time divisions of the Precambrian and early Paleozoic. The "Secondary" eventually was replaced by the late Paleozoic and the Mesozoic. But Arduino's old term *Tertiary* persisted for over two hundred years, and still survives in modern time scales. In its modern conception, the Tertiary makes up all of the Cenozoic (the last 65 million years) except for the Ice Ages that began about two million years ago, for which the term Quaternary is generally in use (figure 2.2). Some authors prefer to subdivide the Cenozoic into two more equal subdivisions: the Paleogene (from 65 to 24 million years ago) and the Neogene (24 million years ago through the present).

Although this latter scheme may seem nicely balanced, the most fundamental break in Cenozoic history occurred between the middle and late Eocene, not at the formal boundary of any of the Cenozoic epochs. If we were to subdivide the Cenozoic into an early Cenozoic "greenhouse" world and a post-middle Eocene "icehouse" world, perhaps some new terms are in

39

order. In a moment of jest at the 1989 Penrose Conference in South Dakota, Spencer Lucas suggested labeling the period from the late Eocene to present the "Toadstoolian," and the Paleocene to middle Eocene the "Eotoadstoolian." Undoubtedly his choice of names was influenced by the whimsical sound of "Toadstoolian," although he could legitimately claim that it was based on the well-known upper Eocene and lower Oligocene exposures at Toadstool Park in western Nebraska. He suggested that if we abandon the mistaken notion of "Terminal Eocene Event," we can replace it with the more accurate "Terminal Eotoadstoolian Event" and retain the popular acronym, "TEE."

Lyell's Percentages

The modern subdivision of the Tertiary into epochs was first suggested by Charles Lyell (figure 2.1) in the third volume of his classic *Principles of Geology* (1833). In this book, Lyell brought geology into the modern era by effectively arguing against supernatural, biblical explanations in favor of natural processes that we see operating around us today. In his third volume, he reviewed the geological record as it was known in his time and created new names to refer to the many rock units recognized as Tertiary. He proposed a fourfold division of the Cenozoic into *Eocene* ("dawn of the recent"), *Miocene* ("less recent"), *Older* and *Newer Pliocene* ("more recent").

But Lyell did not base his conception of Eocene, Miocene, and Pliocene on a particular rock sequence. While traveling in northern Italy in 1828, he had noticed that the thick Tertiary sedimentary sections could be subdivided by their molluscan fossils; older rocks had fewer modern species. Eventually, he followed the work of the French paleontologist Paul Gérard Deshayes, who had studied the molluscs of the Paris Basin. Based on his identifications of 8,000 species from more than 40,000 specimens, Deshayes had also concluded that one could tell the relative age of Tertiary rocks by the abundance of living molluscs.

Martin Rudwick (1978) has shown that Lyell's conception of Eocene, Miocene, and Pliocene was fundamentally different from the conceptual basis of modern stratigraphic units. Instead of basing these time units on a specific rock sequence, Lyell thought of Eocene as the time when about 3% of living molluscs existed; the Miocene was the time when about 19% of molluscs are represented by living species (figure 2.3). Lyell thought that molluscan change was like a ticking clock, with the Eocene as a discrete moment in time along the clock face.

Unfortunately, Lyell's system for the Cenozoic does not work well with the rest of the time scale, which is built on placing boundaries in discrete units based on real rocks and fossils in the field. There are designated *type sections* for most of the subdivisions of the periods of the Paleozoic and Mesozoic. A geologist can actually stand on the outcrop and collect samples or fossils from the internationally agreed "standard." Lyell's conception was not based on a rock unit per se, but on an abstract concept of molluscan change.

There were other practical problems with Lyell's molluscan clock. Molluscan turnover was not a steady, linear process like the ticking of a clock. It was episodic, with periods of rapid evolution and extinction and periods of very little change. Most of Lyell's and Deshayes's "species" cannot be used now, since subsequent taxonomic revisions have raised many of them to generic rank, split off many more new species, and lumped others together. In modern classification, the Lyellian system would be very difficult. Stanley et al. (1980) calculated the modern equivalent of Lyell's percentages. Only 50% of Pliocene (3–5 million years ago) marine snails are modern species, and less than 5% of Miocene (5–24 million years ago) fossil snails are still alive. Lyell defined the Miocene at about 17–19% living species. Virtually no modern species were alive in the Eocene.

When geologists tried to apply the Lyellian "clock" to traditional techniques of measuring stratigraphic sections and plotting ranges of fossils, further problems arose. Lyell's four Tertiary divisions were meant to be chronological, but the were not based on specific outcrops. Paleontologists and stratigraphers, however, were accustomed to designating "stage" names for their local stratigraphic sequences and then attempting to determine how they related to other stages. A sequence of stages became a time scale. Lyell indicated several different areas

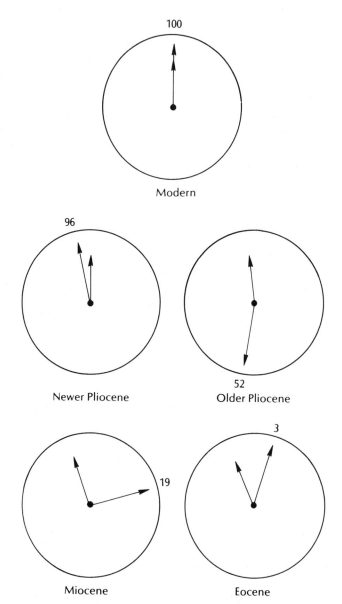

FIGURE 2.3. Lyell's conception of the Cenozoic epochs as "moments" on the "clock" of molluscan turnover. The numbers indicate the percentage of living species in each epoch. (Modified from Rudwick 1978.)

that he thought were "typical" of each epoch, and subsequent authors argued endlessly about which one of these should be designated "the type." Lyell's original conception was based on the Apennine Mountains of Italy, but Deshayes' collections were from the Paris Basin. Detailed study has shown that a few of Lyell's "typical" areas do not belong in his assigned epoch at all but are much younger or older. Finally, the most serious problem is that most of the Paris Basin molluscs are restricted to that region and cannot be recognized elsewhere, so it is impossible to correlate with any other region.

The confusion engendered by Lyell's method is demonstrated by the further subdivision of the time scale. In the Paris Basin, the "upper Eocene" was poorly fossiliferous and had been labeled "lower Miocene" by some geologists. In 1854 Heinrich Ernst von Beyrich coined the term *Oligocene* ("few recent"; there were fewer recent fossils than in the Miocene) for a sequence of rocks in northern Germany and Belgium that was more fossiliferous and apparently younger than the "upper Eocene" of the Paris Basin. The fossils of von Beyrich's Oligocene were more advanced than those of the French Eocene, but not as modern as those of the Miocene, so this epoch was placed between Lyell's Eocene and Miocene. Unfortunately, because the "type Oligocene" is in a different basin and does not overlie the "type Eocene," it is hard to decide where one ends and the other begins. Worse, the "type section" for the Eocene/Oligocene boundary could not be designated in either locality.

The origin of the term *Paleocene* was similarly confusing. In 1874 paleobotanist W. P. Schimper recognized a series of floras in the Paris Basin that he felt were distinct from Lyell's Eocene; he called these "Paleocene" ("ancient recent"). Unfortunately, fossil plants are relatively rare and very difficult to correlate around the world, so the term *Paleocene* did not catch on until the mammals and marine molluscs had also been studied and compared. Most works published early in the twentieth century still used "lower Eocene" instead of Paleocene, so the reader must be careful in interpreting their meaning.

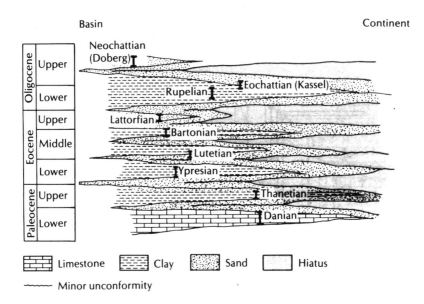

FIGURE 2.4. Depositional history of the type areas of the Paleogene stages and ages in northwestern Europe. Note that the incomplete section is composed of an irregular pattern of transgressive and regressive shallow marine sequences with major gaps or hiatuses (shaded pattern). Each of the stratotype sections is shown by the short vertical bar, and represents only a small portion of Cenozoic time, with major gaps between them. The endemism of the molluscs further complicated intercontinental correlation of these stratotypes. Eventually deep-marine sequences were used as a more complete standard, and the biostratigraphy is now based on microfossils. (Modified from Hardenbol and Berggren 1978.)

Microfossils and Modern Time Scales

Lyell's epochs were based on fossil molluscs that could not be recognized outside the Paris Basin. How do we decide whether a rock in Utah, for example, is Paleocene or Eocene? Even within Western Europe the situation was confused. As reviewed by William Berggren (1971), paleontologists designated type sections for more and more local stages, and soon the arguments focused on whether a particular stage was sequential to another or partially overlapped it. In many cases, two successive stages were not in the same basin, or did not lie in superposition, so it was impossible to determine if they were successive or overlapping. In other cases, it became clear that existing stages left a gap between them which had no name yet. Arguments then followed over whether to extend the upper stage down, the lower stage up, or name yet another stage in between.

The underlying reason for the whole controversy was that Tertiary stages were based upon local sequences within European basins which had undergone complex histories of marine transgression and regression that differed markedly from place to place (figure 2.4). The type section for each stage was part of a time-transgressive cycle that was separated from earlier and later cycles by large unconformities. When geologists tried to trace the stages across basins, they ran into problems. Some stages "pinched out," and others changed into nearshore or non-marine facies which had no diagnostic fossils. In other areas, it became clear that the deep-water sequences were much more complete than the laterally equivalent shallow-marine type areas of a given stage, so that the type area was missing some increment of geologic history (figure 2.4). In short, the stratigraphic terminology was a chaotic patchwork, generating much heated argument but little light.

The solution came in the 1960s and 1970s, when micropaleontologists began to work on deep marine sequences. By its very nature, the deep ocean produces a much more complete stratigraphic record than do shallow marine or non-marine sediments. The deep sea is below the level of most erosion, so there is less chance of unconformities. In addition, much of the deep ocean is covered by a steady rain of shells of dead micro-

45

organisms, which produces a layer-cake blanket of sediment that is relatively complete and undisturbed. By contrast, the shallow marine and nearshore environment is repeatedly eroded each time sea level drops. Nonmarine sediments have even less chance of preservation or completeness, since they are well above erosional base level and are normally doomed to destruction unless they sink into a deep basin and are buried.

As the Deep-Sea Drilling Project (DSDP) proceeded to drill the ocean floor around the world in the 1970s and 1980s, more and more oceanic cores were recovered, producing relatively complete records of the Cenozoic. These cores were full of millions of tiny microfossils, which evolved rapidly through geologic time. Planktonic microfossils were particularly useful, since they had floated in the open ocean and could be found all over the world at a given time. This made correlation from ocean to ocean much more likely. Because the evolution and overturn of microfossils is largely controlled by global climatic changes and their effect on the oceanic water masses in which the organisms lived, these changes were globally synchronous. Eventually, DSDP cores and uplifted marine sediments in New Zealand and Italy became the standards for the Cenozoic. The first or last appearance of specific microfossils were the most practical global markers of Cenozoic time. We now base all Cenozoic time scales on biostratigraphic zones of planktonic foraminiferans and other microfossils.

Micropaleontologists were thus in the key position to resolve the fruitless decades of argument among European stratigraphers. Led by William Berggren, micropaleontologists examined samples of most of the classic type areas to determine how they fit into the deep-sea chronology. This was not an easy task because many of the European sections were deposited under very shallow waters. These conditions were not likely to have many open-ocean microfossils, and often the few microfossils present were poorly preserved. Micropaleontologists showed that many long-entrenched names were useless or overlapped other names, so "sacred cows" had to be slaughtered. Even more discouraging was the discovery that most of the European type sections did not cover much of the time they were supposed to represent. There were large time gaps between the type sections. Because the more complete deep-marine record has become the standard, however, it hardly matters that the European types are incomplete.

Decaying Atoms

So far I have focused on the development of the relative time scale: the sequence of geologic events established by superposition and correlation by fossils. But I have not attached numbers to any of these Cenozoic events. How is this done?

Most numerical ages are produced by radioisotopic dating (see Prothero 1990 for a summary). This technique works on the principle that certain radioactive elements are unstable and decay at a known rate through geologic time. For example, potassium-argon (K-Ar) dating relies on the fact that the unstable parent isotope ^{40}K decays to ^{40}Ar, a daughter product. It takes 1.25 billion years (the *half-life* of this particular system) for half of the parent atoms to decay into daughter isotopes. After one half-life, only half of the ^{40}K remains; after two half-lives, only one quarter; after three half-lives, only one eighth, and so on. Since the half-life and the decay constant are known, one can calculate a numerical age if one can measure the number of parent and daughter atoms.

Using a mass spectrometer, geochemists extract all the parent and daughter isotopes from a sample and make this calculation. In principle this is simple, but in reality it is complicated by numerous problems. The biggest problem occurs when one of the isotopes leaks out of the crystal, so the parent/daughter ratio is altered. In other cases, crystals can become contaminated with too much of either the parent or daughter isotope, again altering the ratio. Although these situations have caused much grief, over the years geochronologists have developed methods and standards to double-check and minimize problems. Nevertheless, there is always the possibility that undetected problems have caused erroneous dates, even with flawless lab procedure.

Another consideration is the analytical error inherent in trying to measure tiny amounts of an element in the mass spectrometer. Even with the best machines, the measurement cannot be repeated exactly, so its precision or reproducibility is assigned an error estimate (usually expressed as "plus or minus" a certain number of years). Even under the best conditions, the error typically ranges between 0.2 and 2 percent. A date of 100 ± 2 million years is actually saying that the true age probably lies between 98 and 102 million years. Error estimates on ages of Precambrian rocks are typically in the order of ±20 million years. For the Eocene and Oligocene, most K-Ar dates have error estimates of ±700,000 years, although the newest techniques of $^{40}Ar/^{39}Ar$ dating produce errors of about 100,000-200,000 years. Fission-track dating of Eocene and Oligocene rocks, on the other hand, typically produces errors of greater than a million years.

The most important thing to remember, however, is that radioactive dating is relevant only in rocks that contain crystals that crystallized from a melt. As a crystal cools, it locks in all the unstable parent atoms in the melt, and the eventual daughter atoms will remain locked into the crystal if it does not leak. The date given is a measure of the time of cooling of a hot igneous or metamorphic rock. Unfortunately, most of the record of geologic history is contained in sedimentary rocks, and these do not form from cooling, but from particles that weather out of pre-existing rock. Individual crystals out of a sandstone could be dated, but this would only date the igneous or metamorphic rock which was the source of the grain, not the sedimentary rock itself.

Consequently, *most sedimentary rocks cannot be directly dated*. Only indirect methods can be used for most sedimentary sequences. This usually means finding an intruded dike, and determining its age relative to the rocks it intrudes. Two dikes of different ages might bracket the age of a rock sequence, although not with much precision. Under the best circumstances, there are datable lava flows or volcanic ash falls interlayered with the sedimentary sequence, and these can be precisely tied to the beds above and below them. For over eighty years, stratigraphers and geochronologists have been looking for places where datable rocks occur in some useful relationship to fossiliferous sedimentary sequences, so that numerical ages can be attached to the relative time scale. In all those years, the time scale has improved considerably, but new dates are always needed for further refinement. For example, the Silurian/Devonian boundary is still only crudely calibrated, and estimates of its age range from 400 to 411 million years (Prothero 1990:282).

In the case of the Eocene-Oligocene transition, finding datable rocks has been particularly difficult. Most of the time scale is based on marine stages, but fresh volcanic ashes or lava flows rarely survive in the marine environment. Wherever possible, these were used (Berggren 1971; Hardenbol and Berggren 1978; Berggren et al. 1985), but they were few and far between. Instead, most of the early time scales had to settle for a green clay mineral known as glauconite, which forms diagenetically on the sea floor under shallow marine conditions. Glauconites were abundant in some of the important European marine sections, so they were widely used for K-Ar dating. Unfortunately, glauconites have all sorts of problems not found in minerals that lock in their isotopes during cooling from high temperatures. Glauconites are notorious for leaking argon gas, making their apparent ages too young. For this reason, most of the early dates from glauconite had to be rejected, and only more recent dates produced under proper lab procedure adequately address the problem. Other early glauconite dates had to be rejected because the equipment was not properly calibrated to screen out contamination from atmospheric argon. Even the freshest glauconite crystals, run under ideal laboratory conditions, can present difficulties (Obradovich 1988). Because glauconites form diagenetically on the sea floor, they can crystallize long after they are buried, and thus lock in a much younger apparent age. They also occur as rounded pellets (replacing fecal pellets of marine organisms), and these can roll around easily and can be reworked from older rocks into younger sediments. This would produce dates which are too old.

For lack of a better choice, however, glauconites were used in nearly every Paleogene time scale. The time scales generated by French geochemist Gilles Odin (Odin 1978, 1982) were heavily dependent on glauconites, and gave ages that were suspiciously young compared to time scales calculated with higher-temperature minerals (e.g., Hardenbol and Berggren

1978; Berggren et al. 1978, 1985). This disparity was intolerably great, even for the Eocene-Oligocene transition. For example, Odin (Curry and Odin 1982; Odin and Curry 1985) placed the Eocene/Oligocene boundary at 32 million years ago, but Berggren et al. (1985) estimated it at 36.5 million years. A difference of over 4 million years is not trivial for events that occurred only 30 to 40 million years ago. The duration of most Paleogene stages in their entirety is less than 4 million years.

Magnetic Reversals

Radioactive decay is not the only steady, unidirectional process that can be used to get numerical ages. Since 1963 it has been known that oceanic crustal rocks are continually erupted at mid-ocean ridges. This oceanic crust then spreads away from the ridge as the crust is pulled apart by convection in the mantle. Sea-floor spreading is one of the driving forces of plate tectonics, our modern theory of the way the earth works. In addition to contributing new oceanic crust, however, sea-floor spreading has other consequences. As the crust pulls apart and new magma wells up along the ridge to fill the gap, the magma cools and become basaltic rock. In that basalt are tiny crystals of iron oxide in the form of the mineral magnetite, Fe_3O_4. These magnetic crystals lock in the direction of the earth's magnetic field at the moment they cool below their Curie point, which is about 580°C (over 1000°F). Thus, sequences of rock spreading outward on both sides of a mid-ocean ridge record the changing history of the earth's magnetic field at the time of eruption.

Since the late 1950s, we have known that the earth's magnetic field has not always maintained its modern orientation (known as "normal" polarity). About 780,000 years ago, a compass needle would have pointed south. This is known as reversed polarity, and the earth's field has flipped back and forth between reversed and normal polarities hundreds of times in the last 65 million years. In 1963 Fred Vine and Drummond Matthews were examining magnetic profiles for the Mid-Atlantic Ridge just south of Iceland. These magnetic profiles were generated by ships towing a magnetometer in the ocean, traveling back and forth in long tracks to generate many parallel

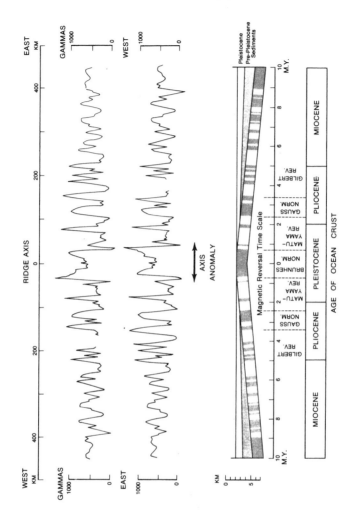

FIGURE 2.5. The discovery of seafloor spreading was originally based on magnetic profiles across the mid-Atlantic ridge south of Iceland. When a magnetometer was towed back and forth across the ridge, the magnetic anomalies alternated between bands which were stronger than the ambient field (positive anomalies, shown with peaks on the magnetic profiles), and bands which were weaker than average (negative anomalies, between the positive peaks). The proof of seafloor spreading came from the symmetry of the profiles (note how the peaks and valleys on the upper profile and the lower profile line up, since they are the same profile flipped over). (From Dott and Prothero 1994; by permission of McGraw-Hill, Inc.)

records across the ridge. When the ship moved over normally magnetized oceanic crust, the magnetic directions of the crustal rock reinforced earth's present magnetic field, and the results were stronger than normal. This is called a positive anomaly. When the ship traversed reversely magnetized crust, however, the rock direction partially canceled out the earth's field, and the result was weaker than normal. This is known as a negative anomaly. Each magnetic profile (figure 2.5) looked like a series of wavy lines, with positive and negative anomalies above and below a baseline representing the intensity of the modern earth's magnetic field.

Vine and Matthews examined many of these "wiggly lines" and profiles, and realized that each was symmetrical. The profile generated to the east of the ridge was the mirror image of the profile generated to the west. The only conceivable explanation for this symmetry was that both sides had been generated by the same process at the ridge, and were now passively spreading away from their source. Since Vine and Matthews first proposed the seafloor spreading hypothesis, we have come to realize that the best record of the earth's magnetic reversals is indeed recorded in spreading seafloor. Each ridge is generating a steady sheet of basaltic ocean floor, which then sinks away from the ridge on both sides. This process has been likened to a magnetic tape recorder, with the "tape" picking up a signal when the ridge erupts, then rolling away from the "tape head" (ridge) and carrying its record with it.

How do we calibrate this "tape recording"? The first test of the seafloor spreading hypothesis was Leg 3 of the DSDP (after two initial "shakedown" cruises), the very first use of the DSDP drilling vessel *Glomar Challenger* as a scientific tool. This cruise drilled sediment cores on both sides of the Mid-Atlantic Ridge in the South Atlantic, and found that microfossils on the seafloor became progressively older away from the ridge. Unfortunately, the seafloor basalt was too altered by exposure to seawater for radiometric dating, and this has generally proven true of seafloor basalts. Paleomagnetists therefore depend on less altered terrestrial volcanics for dating the magnetic reversals of the last five million years of earth history. Enough of these combinations of date and polarity were available to generate a fairly complete magnetic history (Cox et al. 1963). For older periods, the subaerial spreading of the Mid-Atlantic Ridge in Iceland has provided datable magnetics back to 13 million years ago (Mc-Dougall 1977; Harrison et al. 1979). Beyond this point, however, the magnetic pattern of the seafloor cannot be directly dated, so other methods must be used.

One approach takes spreading profiles and extrapolates back from the last 4 million years of dated magnetic history. This was first attempted in 1968 by Jim Heirtzler, Ellen Herron, Walter Pitman, and Xavier le Pichon at Lamont-Doherty Geological Observatory in New York. Heirtzler et al. (1968) compared spreading profiles of the South Atlantic, South Pacific, North Pacific, and Indian Oceans, and found that the South Atlantic had the least variation in spreading history extending back to about 80 million years ago. From the South Atlantic data, they generated a magnetic polarity time scale, which predicted the ages of all the polarity changes since the late Cretaceous. Considering the lack of control points, this extrapolation was a gutsy move. In retrospect, they turned out to be right. Subsequent studies have shown that their initial calculations came within 10% of the currently accepted spreading history. This suggests that the spreading of the seafloor is a remarkably constant process, and its clocklike nature makes it a reliable means of calibrating geological time.

The lack of seafloor control points prior to 4 million years ago was the chief limitation of the Heirtzler et al. (1968) time scale, but in the last twenty years, many revisions have been published utilizing newly available control points. For example, Walter Alvarez, William Lowrie, and others (1977) studied a spectacular sequence of deep marine limestones near Gubbio, in the central Apennines (figure 2.6). These rocks contained marine microfossils from the Cretaceous through the Miocene, and they could also be analyzed magnetically. Although limestones (and other sedimentary rocks) are not erupted and cooled like magnetic lavas, they do contain tiny detrital grains of iron oxides that align themselves with the prevailing magnetic field of the earth at the time they are deposited. This is known as detrital remanent magnetization, and it is the major cause of magnetization of sedimentary rocks. Alvarez, Lowrie,

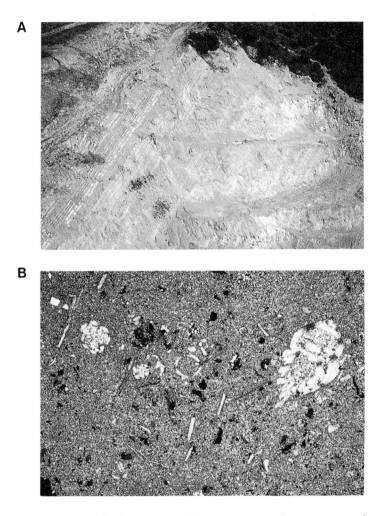

FIGURE 2.6. Uplifted sections of deep-marine sediments proved to be the most practical standards for the Cenozoic time scale, and for dating and magnetic stratigraphy. (A) The Gubbio section in the Italian Apennines was first analyzed in the late 1970s. It contains the Eocene/ Oligocene boundary and the Oligocene/Miocene boundary (Oligocene Scaglia Cinerea is the light colored unit in the middle; dark colored beds on the right are Eocene, and on the left are Miocene). (B) As seen in this photomicrograph, the Gubbio section contained not only abundant microfossils (light-colored shell cross-sections), but also datable biotite grains (long straight crystals) erupted from nearby volcanoes. (Photos courtesy A. Montanari.)

and their colleagues thus generated a distinctive magnetic sequence in sediments that could also be dated by fossils. And they pinpointed the Cretaceous-Tertiary (K/T) boundary, which was also recognized in the spreading profiles. Since dates of 65 to 66 million years had been measured for the K/T boundary on land, Alvarez et al. (1977) could use the K/T tiepoint to pin down the older end of the magnetic time scale in the Gubbio limestones, thereby making it possible to interpolate all the dates between 65 and 13 million years.

About the same time, the decay constants that had been in use for all K-Ar dating were revised, and all K-Ar dates had to be recalculated. Ness et al. (1980) revised the magnetic time scale using these new constants but did not go back to the original dates to recalculate them. In 1981 Lowrie and Alvarez published a new time scale, based on further studies of the Gubbio sequence (Lowrie et al. 1982). Unfortunately, their time scale assumed that the age estimates of the biostratigraphic boundaries were more accurate than the assumption of constant seafloor spreading. Consequently, they suggested that seafloor spreading rates had fluctuated wildly over time. That conclusion goes against everything we know of spreading processes. The problem lay in their assumption that the biostratigraphic boundaries were accurately dated; most biostratigraphers knew that this was a questionable assumption.

By this time, both the Gubbio section (Lowrie et al. 1982) and DSDP Leg 73 in the South Atlantic (Poore et al. 1982) had produced complete marine records with good magnetics and microfossils. What was needed, however, was an association of good high-temperature dates with a distinctive magnetic record to pin down the middle of the magnetic time scale. Between the Heirtzler et al. (1968), Ness et al. (1980), and Lowrie and Alvarez (1981) versions, the Paleogene was oscillating back and forth between the endpoints like a Slinky.

In 1978 I began my Ph.D. thesis work on the magnetics and mammalian stratigraphy of Oligocene deposits of the White River Group in South Dakota, Nebraska, Wyoming, and adjacent areas. The most promising section was a 200 meter (700 foot) sequence of siltstones at Flagstaff Rim, west of Casper, Wyoming (figure 2.7). It had already produced K-Ar dates

FIGURE 2.7. The key section at Flagstaff Rim, near Casper, Wyoming, which provided 700 feet of section with Chadronian mammals and datable ashes. The two light-colored bands above the heads of Malcolm McKenna (left) and John Flynn are ashes F and G; ashes A–D are in the gullies below; ash J is at the very top of the light-colored exposures. See figure 2.8.

estimated an age of 34 million years for the Eocene/Oligocene boundary, based on glauconites from the Castle Hayne Limestone in North Carolina. However, Berggren and Aubry (1983) and Berggren et al. (1985) showed that the samples were misplaced stratigraphically. When Harris and Fullagar (1989) dated a volcanic ash, rather than glauconite, they also got estimates that suggested an age of 34 million years.

Glass and Crosbie (1982) estimated the age of the Eocene/Oligocene boundary at 32.3 million years ago by using glassy droplets of impact ejecta, known as microtektites, from the Caribbean. However, the stratigraphic position of their samples was well below the boundary, so they had to extrapolate to get their estimate. The biostratigraphic assignment of their samples was also problematic. When the same microtektites were re-dated with ^{40}Ar/^{39}Ar dating (Glass et al. 1986), they produced a boundary estimate of 34.4 million years, again by extrapolation.

The most important dates, however, came from the discovery that the deep marine Gubbio sequence contained thin layers of datable volcanic ash (figure 2.6B). For the first time, numerical dates could be directly tied to marine microfossils that define the stages of the Eocene and Oligocene. The first attempt (Montanari et al. 1985) placed the boundary at 35.7 million years, although there were some acknowledged problems with the late Eocene dates. A few years later Montanari et al. (1988) redated their samples from Gubbio, and from another section at Massignano on the Adriatic Coast of Italy. This time they got a boundary estimate of 33.7 million years. With all these dates clustering around 34 million years, clearly something was wrong with the 36.5 estimate of Berggren et al. (1985), but no one could pinpoint the problem.

The solution came when a new dating technology was applied to the old K-Ar dates used by nearly everyone for twenty-five years. Known as ^{40}Ar/^{39}Ar dating, this method uses ^{39}Ar as proxy for ^{40}K, but otherwise it is part of the same K-Ar decay process. The big advantage is that both isotopes are argon gas, so they can be extracted and measured simultaneously. K-Ar methods were hampered by the fact that potassium is solid, not gas, so the process required two different measuring steps. In

K-Ar dating, many different crystals had to be crushed together to get enough material, which tended to mix altered and fresh crystals. In ^{40}Ar/^{39}Ar dating, a single crystal is heated in a vacuum, and it gives off argon gas (both from the parent potassium and from the daughter decay product) first from its edge, and then from the less altered interior of the crystal. The stepwise heating ^{40}Ar/^{39}Ar method compares the age determined from the altered edge with the center, and thus screens out any alteration or contamination. Another method takes each individual crystal and zaps it with a laser, releasing its argon. In this laser-fusion method, the geochronologist can measure dozens of individual crystals and analyze their results to see if they cluster, or if they are so scattered that contamination or leakage must be a problem. Either way, ^{40}Ar/^{39}Ar provided something that K-Ar dating could not: a method of measuring individual crystals, so that contamination could be detected, and error estimates reduced.

When Carl Swisher of the Institute of Human Origins in Berkeley began to redate the Flagstaff Rim ashes in 1989, he discovered something shocking. Many of the K-Ar dates first run by Jack Evernden and Garniss Curtis in 1963 were drawn from contaminated samples. These dates (Evernden et al. 1964) had served as the basis for dating the North American mammalian chronology for over a quarter century, and everyone relied on them (Woodburne 1987). In some cases, the dates were off by as much as 2 million years. Flagstaff Rim Ash J, for example, had been K-Ar dated at 32.5 million years, but laser-fusion ^{40}Ar/^{39}Ar methods gave a date of 34.4 (figure 2.8). Naturally, this radically changed the accepted interpretation that Flagstaff Rim spanned 4 to 5 million years, and also our interpretation that the long reversed interval between Ashes D and J was Chron C12R. After Swisher had completed the revised dating, and I had rerun much of the magnetics (Swisher and Prothero 1990; Prothero and Swisher 1992), it was clear that Flagstaff Rim was a very short sequence, spanning the interval from 34.4 to about 36.5 to 37.0 million years (figure 2.8). There was no way the reversed interval could represent Chron C12R; we now think it represents Chron C15R (Prothero and Swisher 1992).

FIGURE 2.9. Exposure of the volcanic ash (light band about halfway up cliff) known as the "PWL," "Persistent White Layer", "Purplish White Layer" or "5a tuff" south of Douglas, Wyoming (Evanoff et al. 1992). It has been $^{40}Ar/^{39}Ar$ dated at 33.9 million years (Prothero and Swisher 1992), and lies just a few meters below the Chadronian/Orellan boundary (Eocene/Oligocene boundary). The PWL is a widespread marker horizon for the Chadronian-Orellan transition all over eastern Wyoming and western Nebraska.

Fortunately, $^{40}Ar/^{39}Ar$ dating had not only corrected the error at Flagstaff Rim, but also provided the solution to the mystery of the missing Chron C12R. The younger sections of the White River Group in Wyoming, Nebraska, and South Dakota had a long reversed interval during the late Orellan and early Whitneyan North American land mammal "ages" (figure 2.8). In my dissertation, I had difficulty squeezing all this section into the relatively short Chron C10R, since its relative length was consistent all over the White River Group, regardless of the local sediment accumulation rate. There were also volcanic ashes in these sections, but K-Ar dating did not yield usable results. When Carl Swisher ran these same ashes by $^{40}Ar/^{39}Ar$ methods, he was able to obtain reliable dates for the first time. A widespread ash (variously known as the "PWL," "Persistent White Layer," "Purplish White Layer," "5a tuff," and the "100-foot white layer") occurred widely in eastern Wyoming and western Nebraska (figure 2.9). It had long been used as a marker for the contact between the Chadron and Brule Forma-tions of the White River Group, and was close to the boundary between the Chadronian and Orellan land mammal "ages." There were also volcanic ashes in the Whitney Member of the Brule Formation in Nebraska that could not be dated by K-Ar. Swisher's $^{40}Ar/^{39}Ar$ methods produced a date of 33.9 million years on the PWL, and 30.8 million years on the Upper Whitney Ash—which showed that the reversed interval between them spanned almost 3 million years. Here was the mysteriously missing Chron C12R!

When we published our adjustments to the North American dating and magnetics (Swisher and Prothero 1990), it forced a re-evaluation of the Berggren et al. (1985) time scale. Berggren, Kent, Obradovich, and Swisher (1992) have now factored in all the new data mentioned above, and their current estimate of the boundary is about 33.7 to 34 million years. This date is in good agreement with the dates calculated from the Castle Hayne Limestone, the Caribbean microtektites, and Gubbio, although it is not as young as those derived from Odin's glauconites. It seems as though 34 is a compromise between the 36.5 and 32 million year extremes, but it is not a political choice; it is based on new and better data. The striking convergence of so many dates from so many regions suggests that the major conflicts are over, and that this age estimate has probably stabilized. Now that the Eocene/Oligocene boundary dispute seems set-tled, the entire time scale is undergoing further revision as more and more old K-Ar dates are redated by $^{40}Ar/^{39}Ar$ meth-ods (Berggren et al. 1992; Cande and Kent 1992). The chronol-ogy in this book reflects the current state of the time scale, and all the radical changes that it implies for our old conceptions of Eocene-Oligocene history.

The Terrestrial Record

As long as shallow marine fossils occur in a sedimentary sequence, there is some possibility that they can be matched with other marine records that can eventually be correlated with the deep marine microfossil zones. Because the Eocene and Oligocene were defined in marine strata in Europe, they are the reference standard for any use of these terms. However, correlating terrestrial sediments with the marine record is difficult. For most of the later Eocene and the entire Oligocene, there are few places where fossiliferous marine and nonmarine beds interfinger. Most of the fossiliferous terrestrial sediments, such as those of the White River Group, were deposited in the continental interior, far from the areas of marine deposition. Consequently, most methods of correlating land sequences to the marine record are very indirect. The few instances of marine-terrestrial interfingering give some tiepoints, but long intervals of the terrestrial record are unconstrained. This is especially true in North America and Asia, where fossils are almost entirely from the continental interior. In Europe, some shallow marine beds produce fossil land mammals, but the best collections come from collapsed limestone caverns and fissures in the Quercy region of France, which are isolated holes in the ground that cannot be tied to anything.

Consequently, terrestrial stratigraphers have had to develop their own time scales, based on the local sequence of basins bearing land mammal fossils (figures 2.10, 2.11). In North America, this began around the turn of the century (Matthew, 1897; Osborn 1929), but was formalized by a committee headed by Horace Wood (Wood et al. 1941). The Wood Committee set up a series of provincial land mammal "ages," named after regions that produced characteristic fossils of each time period (figure 2.12). However, they did not follow the rules of biostratigraphy. They did not designate local stages based on particular local sections with biostratigraphic ranges of the fossils indicated. Instead, the Wood Committee used a hybrid system of biochronology, determining sequences of fossil faunas independent of biostratigraphic reference sections. Hence, their system did not produce true chronological ages (which must be based on biostratigraphic stages, according to the Code of Stratigraphic Nomenclature). The North American land mammal "ages" must therefore remain in quotation marks.

For North America, the Wood Committee designated a series of land mammal "ages" which it attempted to correlate to the European standard. The Wasatchian (from the Wasatch Formation in Wyoming) was thought to be early Eocene, the Bridgerian (from the Bridger Formation in southwestern

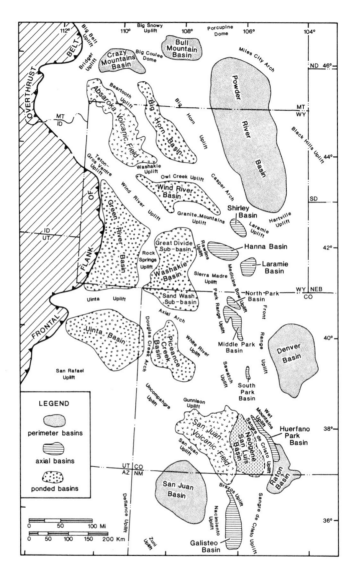

FIGURE 2.10. Location of the major sedimentary basins and their surrounding uplifts in the Rocky Mountain region. Most of these basins accumulated thousands of feet of Paleocene and Eocene terrestrial sediments full of fossil mammals. (From Dickinson et al. 1988; by permission of the Geological Society of America.)

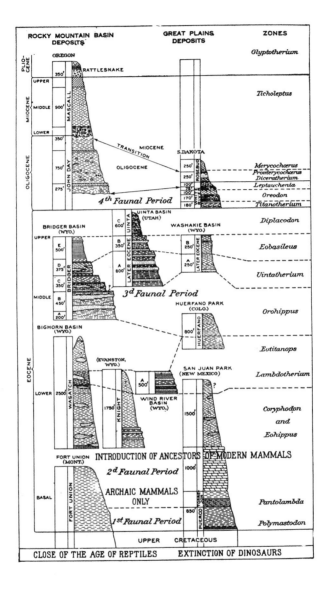

FIGURE 2.11. Osborn's (1909, Fig. 10) conception of the correlation and temporal overlap of the stratigraphic columns in the important Rocky Mountain basins and in the Great Plains. (Courtesy U.S. Geological Survey.)

65

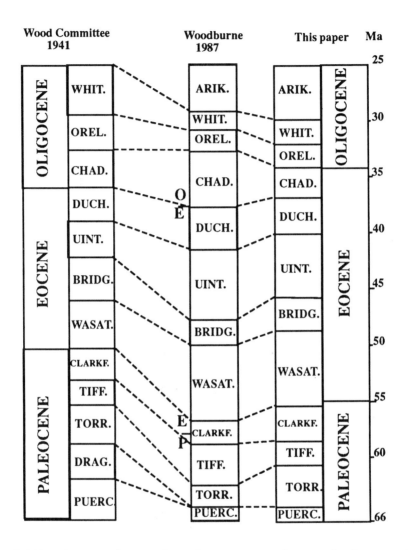

FIGURE 2.12. Evolution of North American Paleogene land mammal "ages" from 1941 to 1993. The Chadronian has moved from early Oligocene to late Eocene, and the Paleocene/Eocene boundary (E/P) has shifted from the middle of the Clarkforkian to the beginning of the Wasatchian. The Duchesnean and Uintan, which were once considered late Eocene, are now correlated with the middle Eocene. The Arikaree an, once considered early Miocene, is now mostly late Oligocene. The Dragonian has been abandoned entirely. Abbreviations: O/E = Eocene/ Oligocene boundary. Full names of North American land mammal ages are shown in figure 2.2.

Wyoming) middle Eocene, and the Uintan and Duchesnean (from the Uinta and Duchesne River Formations of the Uinta Basin in northeast Utah) were considered late Eocene. The Chadronian (from the Chadron Formation of the White River Group) was thought to be early Oligocene, the Orellan (from the Orella Member of the Brule Formation of the White River Group) middle Oligocene, and the Whitneyan (from the Whitney Member of the Brule Formation) late Oligocene (figure 2.13). The Arikareean (from the Arikaree Group in Nebraska) was thought to be early Miocene.

The basis for tying these local land mammal "ages" to the European standard was the common occurrence of certain mammals in both continents. For the early Eocene, we have already seen that there was much migration back and forth across the North Atlantic, so it was relatively easy to correlate the North American Wasatchian with part of the European Sparnacian. By the middle Eocene, however, this faunal interchange had disappeared, so there was little to tie the Bridgerian through Whitneyan to Europe. There were exchanges with Asian mammals during part of this interval, but Asia was isolated from Europe. These difficulties, however, did not deter repeated arguments to decide how the land mammal "ages" related to the global time scale. For example, Wood et al. (1941) thought the Duchesnean was latest Eocene. Scott (1945) thought it was early Oligocene, but Simpson (1946) pushed it back into the Eocene. Wilson (1978) and Emry (1981) questioned whether the Duchesnean could really be distinguished from the Chadronian, and later Wilson (1984, 1986) split the Duchesnean into subages of the Uintan and Chadronian. The trend has now swung back toward recognition of the distinctiveness of the Duchesnean (Krishtalka et al. 1987; Kelly 1990; Lucas 1992), although there is no way to determine whether it is Eocene or Oligocene based on land mammals alone.

The only way these problems could be resolved was by further radiometric dating, and by magnetic stratigraphy. Neither of these methods is hampered by the endemism of local mammal faunas, or by facies changes. By the 1970s, there were numerous K-Ar dates from the Arikaree Group showing that it was mostly late Oligocene, not early Miocene (Obradovich et al.

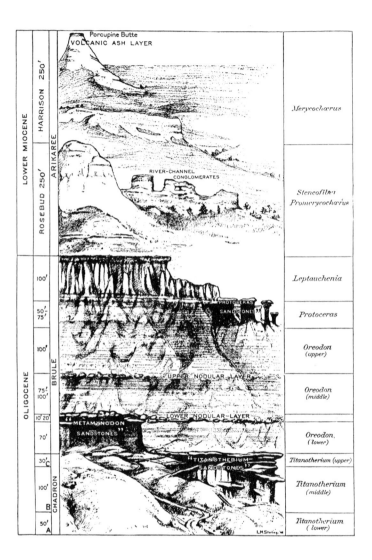

FIGURE 2.13. Osborn's (1929) stratigraphic section through the White River and Arikaree Groups in the High Plains, showing their typical outcrops. Osborn's preliminary mammalian biostratigraphy in shown in the right-hand column. Today the Chadron Formation is considered late Eocene, and the Arikaree Group is mostly late Oligocene. (Courtesy U.S. Geological Survey.)

1973; Tedford et al. 1987). Berggren et al. (1985) concluded that the dates on the Uintan made it middle Eocene, not late Eocene. The most radical change came when Carl Swisher and I reported the new $^{40}Ar/^{39}Ar$ dates on the White River Group (Swisher and Prothero 1990). With the stabilization of the Eocene/Oligocene boundary at 34 million years, there was no question that the new dates made the Chadronian late Eocene, not early Oligocene (figure 2.8). Similarly, the Orellan and Whitneyan became early Oligocene, not middle and late Oligocene.

These changes may seem trivial to the nonspecialist, but to generations of North American vertebrate paleontologists who learned "Chadronian equals early Oligocene" from their very first college class in the subject, this was hard to swallow. It is surprising how many papers are still published that have not come to terms with the current time scale. Some authors try to define it away. For example, there are still those who call the Arikareean the "North American Miocene" to save the concepts they were raised with. Unfortunately, you can't have a "North American Miocene." Lyell's Miocene is defined with reference to a European standard, and its equivalence in North America can only be established by correlation or dating. If the correlation changes, then the European Miocene is still the same, but the North American chronology has to be corrected. This was the whole point of having a discrete sequence of North American land mammal "ages" in the first place! The Wasatchian through Arikareean and their defining mammals are still pretty much the same after over fifty years (Woodburne 1987), but how they tie to the Lyellian epochs in Europe has changed remarkably.

The implications of the changes in the time scale are still not completely resolved. For example, the Asian mammalian sequence has been studied intensively by the Chinese, but only a little radiometric dating or magnetic stratigraphy has been done. Consequently, Asian mammalian faunas have been interpreted with reference to North America, which had a few faunal interchanges with Asia in the late Eocene and Oligocene (Li and Ting 1983; Dashzeveg and Devaytkin 1986; Russell and Zhai 1987; Wang 1992). Carl Swisher and I published the recalibration of the North American sequence in 1990, yet Wang (1992) rejected our conclusions and refused to readjust the Chinese mammalian chronology. When I examined her mammal assemblages in detail (discussed in Berggren and Prothero 1992), I found that the Chinese "early Oligocene" was probably late Eocene (like the Chadronian), and that the Chinese "middle Oligocene" was probably early Oligocene (like the Orellan and Whitneyan). Recently, expeditions to the Paleogene mammal beds of China and Mongolia have recovered samples for radiometric dating, and we shall see if my suggested revisions prove right.

South American land mammal chronology is currently even more confused. We have long known that South America was an island continent through most of the Cenozoic, with its own endemic fauna of pouched mammals, archaic hoofed mammals, and edentates. Not until the late Oligocene did New World monkeys and caviomorph rodents (the group that includes living guinea pigs, capybaras, and chinchillas) enter the continent and diversify. South America remained in "splendid isolation" (in the words of George Gaylord Simpson) until the middle Pliocene, when the Panamanian land bridge arose and allowed the "Great American Interchange" between the Americas. Most of South America's natives were then displaced by immigrant mastodonts, mammoths, camels, horses, sabertooths and other cats from the north. Only ground sloths and armadillos managed to penetrate north to the southern areas of the United States.

In the Eocene, however, South America was just beginning its great diversification of native hoofed mammals and other endemics. The chronology of these South American deposits was originally based on just a few K-Ar dates. The early Eocene Casamayoran faunas were well studied, but the Mustersan (supposedly late Eocene) and Divisaderan (late Eocene or early Oligocene) were much more poorly known and dated. The first appearance of South American monkeys and caviomorph rodents occurred in the Deseadan, which was supposedly early Oligocene. Recently, however, magnetostratigraphy and radiometric dating undertaken by Bruce MacFadden and colleagues (MacFadden et al. 1985) have shown that the

Divisaderan and Deseadan are late Oligocene, spanning roughly 28 to 21 million years ago. This makes them almost 10 million years younger than originally thought. The new dates leave an enormous gap in the South American record from the Mustersan (around 40 million years) to the Divisaderan-Deseadan (around 28 million years). This gap spans most of the Eocene-Oligocene transition that is the focus of this book. Recently, Novacek et al. (1989), Wyss et al. (1990), and Flynn and Wyss (1990) have reported a new fauna from southern Chile, radiometrically dated around 30 million years ago. This may be our first true record of early Oligocene mammals in South America. But the research is still in progress, so it is too early to report their results here. Because the chronology is still poorly known, and the faunas were isolated from the rest of the world, we will not discuss the South American mammal record in this book as much as we cover the other continents.

FIGURE 3.1. Typical landscape during the middle Eocene in North America (Bridgerian or Uintan). A herd of *Orohippus* is running from the multihorned uintatheres in the background. (Painting by Z. Burian.)

The Cooling Begins
The Middle Eocene

Blow, blow, thou winter wind,
Thou art not so unkind
As man's ingratitude;
Thy tooth is not so keen,
Although thy breath be rude . . .

Freeze, freeze, thou bitter sky,
That dost not bite so nigh
As benefits forgot;
Though thou the waters warp,
Thy sting is not so sharp
As friend remember'd not

—WILLIAM SHAKESPEARE, *AS YOU LIKE IT* (1599)

In previous chapters, we have seen how the world experienced maximum warming during the early Eocene. How and when did we make the transition from the warm "greenhouse" world of the early Eocene to the "icehouse" climate we have today? Was it an instantaneous climatic crash, or a long, protracted deterioration?

Fortunately, we have abundant evidence of this climatic transition in both marine and terrestrial sediments. As the dating and correlation of this record becomes better known, we find a surprising fact: the transformation begins much earlier than the "Terminal Eocene Event." The most detailed record of this change can be seen from the chemistry of the oceans as recorded in their sediments.

THE COOLING BEGINS

Messages from Oxygen and Carbon

Oxygen occurs in two stable forms (i.e., not subject to radioactive decay) with different atomic weights. The ^{16}O isotope, with eight protons and eight neutrons, makes up 99.756% of the oxygen on earth. Only 0.205% of the earth's oxygen is the heavier isotope, ^{18}O, with eight protons and ten neutrons. In 1947 geochemist Harold Urey found that the two isotopes are fractionated during evaporation and precipitation of water. A water molecule containing only the lighter isotope, ^{16}O, evaporates more readily than a water molecule made up of the heavier isotope. The lighter water tends to be evaporated from the ocean, leaving the ocean richer in ^{18}O.

In 1954 Urey's student Cesare Emiliani applied this discovery to understanding ancient climates. He knew that microorganisms in the ocean, especially foraminiferans, secrete their calcite shells using the CO_2 from the seawater. For many species, therefore, their calcite has a ratio of oxygen isotopes that reflects the composition of seawater at the time they live. Emiliani measured the oxygen isotope ratios in foraminiferans from a number of deep-sea cores spanning the last ice ages. Planktonic foraminiferans were used, but benthic foraminiferans proved to be better indicators, since they live on the bottom where temperature change is minimal except for long-term climatic changes (Emiliani 1954). Much later, calcareous nannoplankton were also tried, but they were often heavily recrystallized, and apparently they do not secrete their calcite in equilibrium with seawater (Goodney et al. 1980). Eventually, oxygen isotopic analysis of benthic foraminiferans became the primary method of determining ancient oceanic temperatures.

Temperature, however, is not the only factor affecting oxygen isotope ratios. When the earth is in a glacial episode, ice volume is even more important (Shackleton and Opdyke 1973). Isotopically light water evaporates from the ocean and then snows on polar regions, locking large amounts of ^{16}O-rich water in the ice caps. This leaves the ocean depleted in ^{16}O and enriched in ^{18}O during glacial episodes. When the ice caps melt, the ^{16}O returns to the ocean, and the relative amount of ^{18}O decreases. By the late 1970s, most paleoceanographers came to realize that the ice signal was much stronger than the temperature effect, so during times of polar ice caps, we can interpret the oxygen isotope ratios as an ice signal. However, for much of the Mesozoic and Cenozoic, we know there were no significant polar ice caps. During these times, it is likely that temperature was the major factor controlling oxygen isotope ratios.

By convention, oxygen isotope ratios are expressed as ratios of $^{18}O/^{16}O$ (written "$\delta^{18}O$," and pronounced "del O-18"), and they are measured with reference to a laboratory standard. The standard happens to be the extinct squid-like belemnites from the Cretaceous Pee Dee Formation of South Carolina, so it is written "PDB." Since the ratio has ^{18}O in the numerator, increasing values are heavier (indicating cooling and/or ice volume increases); decreasing values are lighter (meaning warming and/or ice volume decreases). The formula for calculating $\delta^{18}O$ produces values that vary only a few parts per thousand (written ‰, or "per mil").

Since Emiliani's pioneering work on the Pleistocene, a number of deep marine sequences have produced oxygen isotopic records for the Cenozoic (Devereaux 1967; Shackleton and Kennett 1975; Savin et al. 1975; Savin 1977). No matter which version of the oxygen isotope curve one examines, it is clear that the oceans were isotopically lightest (i.e., warmest) during the early Eocene (figure 3.2). Buchardt (1978) reported mean temperatures in the North Sea that were downright subtropical—as high as 28°C (82°F). Even the high latitudes were remarkably warm. According to Shackleton (1986) and Miller et al. (1987), oxygen isotopic values from the shells of benthic foraminiferans taken from deep sea cores in the South Atlantic suggest deep oceanic waters were as warm as 12°C (54°F) in the early Eocene; present values are near freezing. [Since the freezing point of salty water is several degrees lower than 0°C (32°F), deep ocean waters can actually get colder than this temperature and still not freeze]. Savin (1977) reported values around 12° or 13°C in the early Eocene for the deep waters of the North Pacific, thus showing that the oceans were unusually warm right to the bottom. Shackleton and Kennett (1975) reported planktonic foraminiferan paleotemperatures as warm as 19°C (66°F). Frakes (1979, fig. 7-2) calculated that oceanic

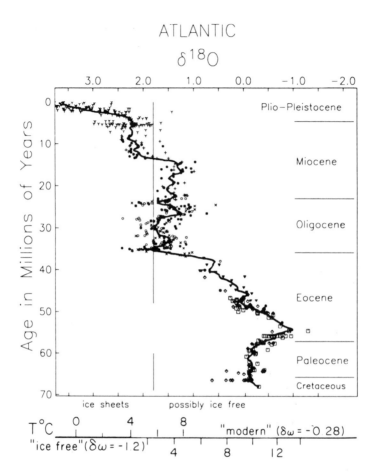

FIGURE 3.2. Oxygen isotope history for the Cenozoic. The temperature scale for the right, "ice free" side (bottom temperature scale) is different from the upper scale used for the glaciated intervals (left side). (Modified from Miller et al. 1987.)

surface temperatures ranged from about 14°C (57°F) near the poles to 22°C (72°F) at the equator.

After the peak of warmth in the early Eocene, all the oxygen isotopic data show a marked increase in isotopic values (Savin 1977). This almost certainly reflects global cooling in the Eocene, although by the Oligocene it was partly caused by ice volume increases as well. In the first significant step, the isotopic value of deep ocean benthic foraminifers increased by about 1.0‰. This occurred somewhere between magnetic Chrons C21N and C22N (Miller 1992), about 47 to 50 million years ago, which is near the early/middle Eocene boundary (within Chron C22N, about 50 million years ago). Since the early Eocene was apparently ice free, this change probably corresponds to a global cooling of about 4 to 5°C (7 to 9°F) (Miller 1992). By contrast, the temperature fluctuated only about 2°C in the deep sea during the most recent ice ages. A second increase of another 1.0‰ occurred near the end of the middle Eocene, suggesting another global cooling of about 4 to 5°C. The timing of this event has not yet been well constrained, but it appears to occur between Chrons C17N and C18N (Miller 1992). The marine middle/late Eocene boundary occurs within Chron C17N (Berggren et al. 1985), which is now dated around 38 to 37 million years ago (Cande and Kent 1992). Similar isotopic ratio increases occurred in planktonic foraminifera, which reflects cooling in surface waters (Shackleton and Kennett 1975). Clearly, the early Eocene greenhouse was deteriorating rapidly during the middle Eocene, yielding deep ocean temperatures of only 5°C (41°F) in the late Eocene (Miller et al. 1987). By then, the North Sea had cooled to about 15°C (60°F) (Buchardt 1987).

Like oxygen, carbon has a number of isotopes. The most familiar is ^{14}C (carbon-14), which is radioactive and decays to nonradioactive ^{14}N. It is used for radiocarbon dating, but the decay rate is so rapid (its half-life is only 5,370 years) that it can only be used for dating events younger than 60,000 years, although it has occasionally been pushed to 80,000 years. The two main stable (i.e., non-radiogenic) isotopes of carbon are ^{12}C (6 protons and 6 neutrons, making up 98.89% of the earth's carbon), and ^{13}C (6 protons and 7 neutrons, making up the rest). Like oxygen isotopes, both of these stable carbon isotopes are found in the CO_2 in seawater and are incorporated into the calcite of marine shells. As with oxygen isotopes, $^{13}C/^{12}C$ ratios are calculated in parts per mil measured against a PDB standard and are expressed in terms of $\delta^{13}C$. Typically, isotopic analyses are done on both the oxygen and carbon isotopes from the same cores, so that their signals can be compared.

Unlike oxygen isotopes, however, carbon isotope ratios do not directly track temperature or ice volume. Instead, they seem to be most sensitive to changes in oceanic circulation. CO_2 molecules that contain the light isotope of carbon are preferentially fractionated during photosynthesis. Organic matter tends to be relatively enriched in ^{12}C and low in ^{13}C. When this dead tissue decays, abundant ^{12}C is released into the water, driving the ratio down toward "light" (enriched in ^{12}C and depleted in ^{13}C) carbon. On the other hand, when "light" organic matter is trapped in the deep ocean, ocean waters are relatively enriched in ^{13}C (i.e., the $\delta^{13}C$ is more positive). During periods of oceanic overturn and upwelling, this light carbon is released from trapped bottom waters and spreads through the ocean, driving the $\delta^{13}C$ negative. The contrast between $\delta^{13}C$ in surface- and bottom-dwelling foraminiferans is also a good indication of the degree of oceanic mixing or stagnation.

So what caused the cooling signals in the middle Eocene? For years, conventional wisdom held that the Antarctic continent had maintained its early Eocene ice-free state well into the middle Miocene, about 12 million years ago. In 1987 glacial deposits overlain by a lava dated at 49.4 million years were reported from King George Island in the Antarctic Peninsula (Birkenmajer 1987; Birkenmajer and Zastawniak 1989). This would be the first evidence of Cenozoic Antarctic glaciers, and its date seems to place it around the beginning of the middle Eocene cooling. Wei (1989) redated ice-rafted sediments from deep-sea cores in the Pacific sector of the Southern Ocean (Margolis and Kennett 1971), which demonstrated that icebergs were floating away from Antarctica in the middle Eocene. However, the King George Island glacial deposits are limited to the high mountains on the northernmost islands of the Antarctic Peninsula. Kennett and Barker (1990) argue that these deposits cannot be used as evidence of full-scale continental glaciers or ice sheets on the Antarctic continent during the Eocene. The floral record (Kemp 1975; Truswell 1983; Case 1988; Mohr 1990) indicates that cool temperate forests (including the southern beech *Nothofagus*, podocarps, araucarias, and abundant ferns) persisted on Antarctica through most of the middle and late Eocene (figure 1.11). Fern spores in particular indicate that there were no long-term frosts. In fact, the *Nothofagus* cool-temperate flora returned to the Antarctic peninsula (including King George Island) after the retreat of the early middle Eocene mountain glaciers. So although global cooling had begun in the middle Eocene, it did not yet lock the poles in ice.

Middle Eocene Marine Life

The middle Eocene cooling in the oceans should have produced an obvious response from marine life. In many marine organisms, the response is clear, but in others, the pattern is not so straightforward. The early-middle Eocene transition gives a puzzling mixture of signals, only some of which indicate cooling. The middle-late Eocene transition, on the other hand, was truly a major extinction event, particularly for warm-water marine species.

Most of the research has focused on foraminiferans, which are abundant in the calcareous sediments of the world's oceans. The changes in both planktonic and benthic foraminiferans during the 12 million years of the middle Eocene were complex (figure 3.3). According to Anne Boersma, Isabella Premoli Silva, and Nick Shackleton (1987), the beginning of the middle Eocene (planktonic foraminiferan Zone P8, about 51 million years ago) was marked by several degrees of cooling in the high latitudes and in bottom waters that flowed from high latitudes all the way to the mid-Atlantic and Gulf of Mexico. However, there was little cooling in the near-surface waters, especially in the equatorial belt. Instead, equatorial warm surface waters expanded poleward, which triggered equatorial upwelling and produced abundant siliceous sediments in the

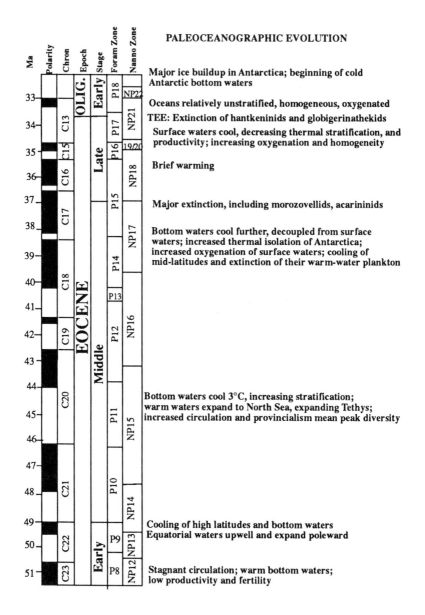

FIGURE 3.3. Summary of marine events in the Eocene (based on Boersma et al. 1987; time scale modified from Cande and Kent 1992 and Berggren et al. 1985.)

Atlantic. Clearly, the stagnant water conditions of the early Eocene were becoming destabilized.

Shortly after the beginning of the middle Eocene (planktonic foraminiferan Zone P11, about 47 million years ago), the Atlantic became more intensely stratified by temperature. Bottom waters and even temperate surface waters cooled by 3°C, creating a significant thermal gradient between surface and bottom. With so many different temperature regimes to occupy, planktonic foraminiferans diversified, with old tropical species persisting along with newly evolved cold-tolerant species. Warm waters bathed the North Sea, allowing Tethyan foraminiferans to spread much further north. The vigorous circulation patterns subdivided oceans into smaller biogeographic subprovinces, and foraminiferans show their highest provincialism at this time. Increased circulation also brought up nutrients from the deeper waters, and from the ocean margins into the center of the oceanic gyres, which are normally nutrient depleted. These nutrients enriched surface waters and led to large-scale plankton blooms. Increased upwelling is also demonstrated by the decreased differerences in carbon isotopes between surface-dwelling and bottom-living foraminiferans.

At the end of the middle Eocene (planktonic foraminiferan Zones P13–P14, about 38 million years ago), the effects of cold bottom waters intensified. Bottom waters were thermally decoupled from low-latitude surface waters, so that the oxygen isotopic histories of surface and deep waters began to follow different paths. More importantly, warm tropical waters were prevented from mingling with polar waters. These in turn became trapped in the Antarctic region, circulating around the continent to further accentuate the thermal isolation of Antarctica, and increasing the gradient of temperatures between pole and equator. Although the tropics did not cool significantly, the temperate regions did. As a result, warm-water foraminiferans disappeared from the mid-latitudes. More importantly, the highest rate of extinction and overturn occurred at the end of Zone P14, with over 18 species of foraminiferans becoming extinct (Boersma et al. 1987). Two typically tropical groups, the

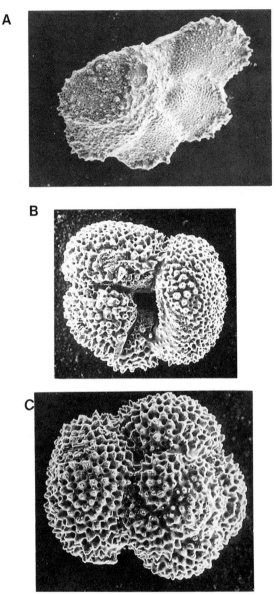

FIGURE 3.4. Typical warm-water planktonic foraminiferans which died out at the end of the middle Eocene. They include (A) *Morozovella lehneri*; (B) *Acarinina bullbrooki*; (C) *Truncorotaloides rohri*. Each specimen is about 300-400 microns in diameter. (Photos courtesy G. Keller.)

morozovellids and acaraninids (figure 3.4), were among the victims.

In addition to these changes in the planktonic and smaller benthic foraminiferans, the disc-shaped nummulitids—which had been abundant across the Tethys—were also affected (figure 1.13). Large nummulitids disappeared completely at the end of the middle Eocene. Only smaller nummulitids managed to survive into the late Eocene and straggle on into the Oligocene.

At the base of the marine food pyramid are the microscopic algae known as coccolithophorids, the dominant group of organisms among the calcareous nannoplankton (nanno- is the prefix for submicroscopic). Members of the yellow-green algae (Chrysophyta), these algae are single-celled and only a few microns in diameter (figure 3.5). They secrete button-shaped calcareous plates, called coccoliths, which surround their cell like overlapping shields. When the plant dies, the coccoliths are shed and dispersed. Since these plants live by the millions in shallow marine waters in the photic zone, they are the primary food of larger microfossils such as foraminiferans. During the Cretaceous, trillions of coccoliths accumulated as thick deposits of a crumbly kind of limestone we know as chalk.

Coccolithophorids are very sensitive to water temperature, and some species are ideal for tracking the change in ocean surface temperatures. Most of the species are also long-ranging. It is common for one species to first appear, or become locally extinct, when the water temperature changes, even though it persists elsewhere (McIntyre and Bé 1967; Haq et al. 1977; Aubry 1992). Biostratigraphers must be careful of first and last occurrences that are time-transgressive, reflecting water temperature changes rather than true extinction or evolution. Coccolithophorids are also sensitive to nutrient changes. But since carbonate and sunlight are normally abundant at the ocean surface, they can thrive even when the ocean becomes stratified and abundant nutrients do not reach the surface (Aubry 1992). The microscopic siliceous algae known as diatoms, on the other hand, must have a supply of scarce silica from upwelling of deep ocean waters to flourish; they would not thrive in a stratified, poorly circulating ocean.

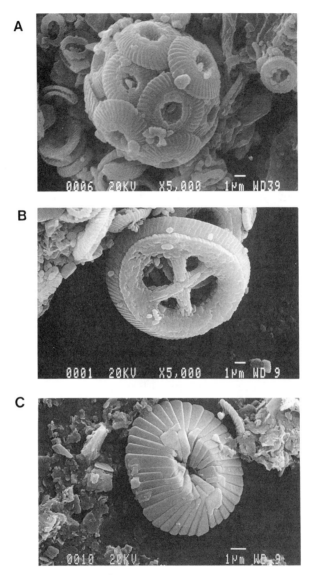

FIGURE 3.5. Coccolithophorids (A) are microscopic single-celled algae which secrete button-shaped calcareous plates called coccoliths (B–C) around themselves in life. These representative Eocene examples include (A) the complete coccosphere of *Cyclicaryolithus floridanus*, and individual coccoliths of (B) *Chiasmolithus altus*; (C) *Coccolithus pelagicus*. Scale bar is one micron. (Photos courtesy Wuchang Wei.)

Keeping these factors in mind, it is not surprising that Aubry (1992) has shown that a major extinction in the calcareous nannoplankton occurred in the high latitudes (but not yet in the tropical regions) during the early-middle Eocene transition. Cores analyzed by Shackleton and Kennett (1975), Shackleton (1986), and Miller (1992) show a deep-water cooling in high-latitudes. The signal is strongest in the southern ocean, although a similar trend has been detected in the north Pacific (Bukry and Snavely 1988). Throughout the middle Eocene, there was continual turnover in lower-latitude nannoplankton, reflecting climatic instability. However, the total diversity of tropical taxa remained high through the middle Eocene. Like the high diversity of middle Eocene foraminiferans (Boersma et al. 1987), this reflects a stratified tropical ocean, which allowed species of different temperature and depth requirements to coexist in the same water column. Aubry (1992) compares this phenomenon to an ecotone, the boundary zone between two different habitats, in which a high diversity of species from both regions coexist where their ranges overlap.

At the end of the middle Eocene, there was a major extinction event of many long-lived species of tropical nannoplankton. Although cooling of tropical waters is partially responsible, Aubry (1992) argues that evidence from planktonic foraminiferans and carbon isotopes show that the stratified ocean was disappearing and becoming more oxygenated at greater depths. This would break up the tropical ecotone, making the water column more homogeneous at depth at the end of the middle Eocene.

The other important microscopic marine algae are the diatoms (figure 3.6). These organisms range from a few tens to a few hundred microns in diameter, and they secrete siliceous skeletons shaped like nested Petri dishes. Diatoms flourish anywhere there is oceanic upwelling to bring scarce silica and other nutrients to surface waters. Today, for example, there are millions of diatoms in upwelling currents around Antarctica, producing thick diatom oozes. During the Miocene, upwelling off the California coast produced enormous blooms of diatoms, which built the thick deposits of diatomite now seen in the middle to late Miocene Monterey Formation. This diatomite is so

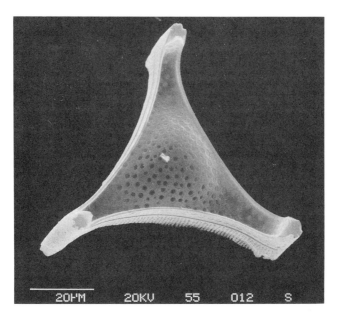

FIGURE 3.6. Diatoms are microscopic single-celled algae which secrete two siliceous capsules which nest in one another like a petri dish. Diatoms can be disc-shaped, lozenge-shaped, or in this case (*Trinacria excavata*) triangular. Scale bar is 20 microns. (Photo courtesy J. Barron.)

voluminous that it is mined commercially for water filters, Kirin beer, and kitty litter.

According to Jouse (1978) and Fenner (1985), the early-middle Eocene transition showed only minor replacement of species typical of the early Paleogene. The middle-late Eocene transition, on the other hand, shows an *increase* in diversity as new species were added to the early-middle Eocene survivors. Baldauf (1992) suggests that the latter represents increasing floral provincialism as the circulation around Antarctica is decoupled from the higher-latitude circulation. If Aubry's (1992) hypotheses are correct, this diversification might also reflect an increase in upwelling as the previously stagnant middle Eocene ocean begins to circulate. In any case, diatoms go against the general trend of extinction at the end of the middle Eocene.

Molluscs were the original basis for Lyell's and Deshayes's conception of the Eocene. How did molluscs respond to middle Eocene events? Thor Hansen (1988, 1992) plotted molluscan diversity in the U.S. Gulf Coast, which has an extensive marine record. A peak of molluscan diversity occurred in the warm, equable early Eocene, in agreement with the trends already discussed. The beginning of the middle Eocene (lower Lisbon Formation) shows a striking decline in diversity of molluscs, especially of gastropods such as the high-spired family Turridae. Bivalves show less of an effect, although some families (such as the tellins) declined severely by the beginning of the middle Eocene. Molluscs then recovered and diversified to even higher levels, reaching their all-time peak late in the middle Eocene (Upper Lisbon–Cook Mountain Formations). The end of the middle Eocene (the contact between the Gosport and Moody's Branch Formation, upper planktonic foram Zone P14) was marked by the most severe decline of all, with 89% of the gastropods and 84% of the bivalves becoming extinct. Another lesser extinction happened in the early late Eocene (between the Moody's Branch and Yazoo Formations, middle planktonic foram Zone P15), wiping out 72% of the gastropods and 63% of the bivalves. After this, almost all molluscan families were at very low diversities in the late Eocene Yazoo Formation and equivalents. Hansen (1988, 1992) shows that these diversity changes are closely correlated with temperature trends documented by Wolfe and Poore (1982), but not with sea level curves or shelf area. Even more revealing are the ecological preferences of these molluscs. Molluscs with demonstrated warm-water or cool-water preferences diversified and declined in almost exact correpondence with the temperature curve. Clearly, cooling was the most influential factor in molluscan decline, and the extinctions at the end of the middle Eocene were the most severe in the Paleogene.

Next to molluscs, the most abundant hard-shelled invertebrates are echinoids. Mike McKinney, Ken McNamara, Burt Carter, and Steve Donovan (1992) analyzed the changes in echinoid diversity during the entire Paleogene. Unfortunately, their data typically include only one point for each stage of the Eocene, so there is not enough resolution to see if there were discrete changes at the beginning or end of the middle Eocene. Nevertheless, it is clear that the early Eocene and early middle Eocene were peaks in echinoid diversity. In some regions (especially the Caribbean) it is possible to see a major decline in echinoids in the middle of the middle Eocene. Unlike most every other marine group, however, there is no major extinction of echinoids at the end of the middle Eocene.

In summary, the 12 million years of the middle Eocene show a long and complex record of cooling and biotic response. The beginning of the middle Eocene was marked by the first decline in the greenhouse conditions that warmed the early Eocene. Sluggish oceanic circulation became destabilized as cold waters began to sink and spread north from the Antarctic. Soon thereafter, the oceans became highly stratified into surface and deeper waters. We see a few extinctions of tropical organisms, but the early middle Eocene is actually a peak in diversity of foraminiferans and coccolithophorids, because the oceans offered a variety of depth- and temperature-stratified habitats, as well as enhanced provinciality. At the end of the middle Eocene, cooling increased dramatically, and even mid-latitudes were affected. Many of the warm-adapted plankton, especially coccolithophorids and foraminiferans, went extinct. The warm tropical Tethyan belt cooled and narrowed, which was fatal to a wide variety of warm-adapted molluscs and benthic foraminiferans, including the large nummulitids. In the marine realm, the end of the middle Eocene—not the end of the late Eocene, as conventionally believed—was the true "mass extinction event." Late Eocene and Oligocene microfossils and molluscs never again recovered to the diversity levels of the late middle Eocene. Only echinoids seem to be an exception to this trend.

Floral Fluctuations

The signal from the oceans is clear. What about the terrestrial record? Fossils of land plants, particularly analyzed according to Wolfe's (1971, 1978, 1990) methods of leaf margin analysis, provide the most sensitive data. Although land plants do not yield as many data points as marine fossils, or as fine a resolution as deep-sea cores, similar trends can be seen. Wolfe

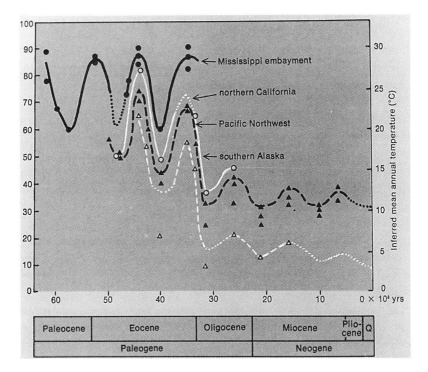

FIGURE 3.7. Paleotemperature curve based on the percentage of entire-margined leaves in radiometrically-dated floras from four different areas of North America. Note the abrupt cooling at the end of the middle Eocene, the late Eocene warming, and the catastrophic cooling in the early Oligocene. (From Wolfe 1978; by permission of *American Scientist*)

(1978) discerns two episodes of cooling in the Eocene of North America (figure 3.7), one at about 50 million years ago (Wasatchian-Bridgerian transition in North American land mammal "ages") and another somewhere around 40 million years ago (approximately the Uintan-Duchesnean transition). The second cooling event is not exactly coincident with the marine middle/late Eocene boundary (which is now dated around 37 million years ago), but the calibration of the land plants is not very precise.

THE COOLING BEGINS

In each cooling episode, mean annual temperature declined from about 30°C (86°F) to about 21°C (70°F) in the U.S. Gulf Coast, and from 23 or 24°C (74 or 76°F) to about 17°C (62°F) in northern California. Continental areas showed a much stronger response to these climatic changes than did the open ocean. There were temperature swings of 7 to 11°C (11 to 16°F) in continental North America compared with only 4 to 5°C (7 to 9°F) of cooling in the deep ocean waters. Another striking difference is that late middle Eocene floras indicate that land temperatures rebounding to levels as warm (30°C) as those of the early Eocene, while oxygen isotopes show the late middle Eocene ocean never quite returned to the warmth of the early Eocene.

This temperature trend can also be seen in the characteristics of the vegetation. In the early middle Eocene, seasonally dry climates must have prevailed in southeastern North America, because the fossil floras contain abundant deciduous plants, including legumes, laurels, oaks, and the walnut family (Wolfe 1986). While early Eocene vegetation was fully moist and tropical over a wide range of latitudes, early middle Eocene floras in the Rocky Mountains show some evidence of cooling and drying (Wing 1987). Part of this trend is related to the development of a series of volcanic uplands (Axelrod 1966; Wing 1987), which fragmented the Rocky Mountain region into floras of different altitudes and average annual precipitation. The early middle Eocene marked the culmination of the Laramide Orogeny, which subdivided the Rocky Mountains into deep basins separated by basement uplifts. Nevertheless, some of the trend recorded in the regional floras is probably a reflection of global climate as well.

For example, the floras from the famous Green River lake beds (figure 3.8) in southern Wyoming and northeastern Utah (figure 2.10) include many species with small, thick-textured leaves, and some species that were clearly deciduous. MacGinitie (1969) called this vegetation "open woodland," although Wolfe (1985) argued that it was more like a semideciduous subtropical to paratropical forest. Many species there are associated with seasonally dry conditions, including poplars and *Cardiospermum* vines. The Boysen flora from the upper

FIGURE 3.8. Thousands of feet of finely laminated middle Eocene lake shales from the Green River Formation, Hell's Hole Canyon, Uinta Basin of Utah.

part of the Wind River Formation (MacGinitie 1969) has a mixture of palms, herbaceous monocots, poplar trees, and vines, suggestive of low-stature vegetation on an open floodplain (Wing 1987). The Kissinger Lakes–Tipperary flora from the Aycross Formation in the western Wind River Basin (MacGinitie 1974) contains ferns, horsetails, conifers, and 44 species of angiosperms; 60% of these are deciduous. Although many species were restricted to subtropical to tropical climates, the high percentage of deciduous vegetation indicates that conditions were too dry to support tropical rainforests. MacGinitie (1974) likened this flora to plants found today at elevations of 1000 meters (3300 feet) on the southwest coast of Mexico.

FIGURE 3.9. (A) Profile of Amethyst Cliff in northeastern Yellowstone National Park, Wyoming, showing the remains of 27 successive middle Eocene fossil forests, each killed and buried by a series of volcanic ash flows, totaling 2000 feet in thickness. (Courtesy U.S. Geological Survey.) (B) Two successive fossil forests buried *in situ*. (Courtesy Yellowstone National Park.)

Even more spectacular is the early middle Eocene sequence of fossil forests in northeastern Yellowstone Park (figure 3.9). On a cliff face called Specimen Ridge, at least 27 successive layers of petrified trees were fossilized as they stood, by repeated eruptions of glowing volcanic ash clouds (Dorf 1960, 1964). The fossils include a mixture of subtropical trees, such as figs, avocados, and bay laurels, along with warm-temperate magnolias, chestnuts, sycamores, walnuts, and many conifers, including pines and sequoias. The abundance of conifers and smaller leaf types seems to indicate somewhat cooler but less seasonally dry climates (Wing 1987). Some geologists have argued that these tree stumps were transported upright by the thick, viscous volcanic mudflows (Fritz 1980, 1986; Coffin 1976). The climatic implications would therefore be suspect because they may represent a transported mixture of vegetation. However, Wing (1987) demonstrates that much of the Yellowstone flora is clearly *in situ*, and that such mixtures of tropical and temperate plants were not unusual in the Bridgerian.

The terrestrial evidence for cooling at the end of the middle Eocene is much more striking. For example, late middle Eocene fossils of the lower Clarno flora in Oregon (figure 3.10), Steels Crossing flora in Washington, and Kushtaka flora in Alaska represent paratropical rainforests, similar to those now found in Taiwan or southern China (Wolfe 1971). They were dominated by broad-leafed evergreen plants, many with drip tips. There were abundant lianas, palm trees, and many members of tropical families such as paw-paw trees, dipterocarps (including many paratropical Asian species), and icacina and moonseed vines. According to Wolfe's (1978, 1990) methods of analysis, these paratropical floras flourished in mean annual temperatures of about 20 to 25°C (68 to 77°F); seasonal shifts in mean temperature, moreover, spanned only 5°C (9°F).

By contrast, the middle Clarno flora of the late Eocene includes a mixture of upland broad-leaved evergreen, deciduous, and coniferous species, including cycads, pines, sequoias, sumacs, poison ivy, birches, sycamores, laurels, plums, blackberries, and wild grapes (Wolfe 1971). According to Wolfe (1978, fig. 4), these plants grew in a climate with a mean annual temperature of 15°C (60°F) and a mean annual range of

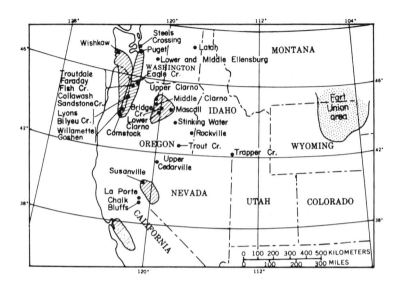

FIGURE 3.10. Location of some Tertiary plant localities mentioned in the text. (Modified from Wolfe 1971.)

temperatures of about 7°C (11°F). In the Puget Group of Washington, the late middle Eocene paratropical Steels Crossing flora was replaced by a late Eocene flora that was very similar to the middle Clarno flora: broad-leaved evergreen and deciduous. In the late middle Eocene of the Gulf of Alaska, broad-leafed evergreen forest was succeeded by conifer-oak-laurel forest (Wolfe 1971), indicating mean annual temperatures of about 12°C (54°F) with an 8°C (11°F) mean annual range of temperature(Wolfe 1978, fig. 4).

   In addition to the megascopic remains of plants, pollen records are also very informative. Frederiksen (1991) analyzed the pollen in middle Eocene marine sequences from both the U.S. Gulf Coast and the San Diego areas. Both regions had paratropical vegetation in the early Eocene, similar to that indicated by fossils further north. But the cooling and drying trend occurs not in the beginning or end of the middle Eocene, but in the middle of the middle Eocene (nannofossil Zone NP16, late

Lutetian, about 42 million years ago). Tropical palms, elms, and walnuts were replaced by sycamore, hickories, and abundant shrubs, herbs and grasses. The drying trend of the middle middle Eocene detected in fossil pollen is confirmed by ancient soil horizons in the San Diego region (Peterson and Abbott 1979), which indicate seasonal, semiarid climates with about 50 to 75 cm (20 to 30 inches) annual rainfall, mostly in the form of occasional flash floods. The dry season was arid ẽnough for crystallization of salt in the soil, and formation of thick caliche horizons. At the end of the middle Eocene, further cooling and drying is indicated by the increase in oak pollen.

In summary, the floral record shows changes at the beginning and the end of the middle Eocene that seem to correspond to the oceanic cooling shown in the oxygen isotope record. The early-middle Eocene transition is represented on land, however, by only slight evidences of cooling and drying. The middle-late Eocene transition, on the other hand, shows significant evidence of cooling of about 7 to 11°C (11 to 15°F) in terrestrial deposits of northwestern North America, and a loss of tropical elements in both Europe and North America. Clearly, by the end of the middle Eocene the tropical rainforests were on the decline around the world.

Bridger, Uinta, and Duchesne River

About 250 miles (360 kilometers) south of the early Eocene beds in the Bighorn Basin are basins that set the standards for middle Eocene vertebrates (figure 3.11). The Bridger Basin in southwestern Wyoming (figure 3.12A) was the basis for the early middle Eocene Bridgerian land mammal "age" (figure 2.11). Just south of the Bridger Basin, on the other side of the Uinta Mountains in Utah, lies the Uinta Basin, which became the classic area for the late middle Eocene Uintan and Duchesnean land mammal "ages." The Uintan takes its name from the Uinta Formation (figure 3.12B), and thick sequence of greenish-gray sandstones and mudstones in the eastern Uinta Basin, that changes into dark gray mudstones and lake shales in the western part of the basin. The Duchesne River Formation, a thick sequence of red sandstones and shales that overlies and

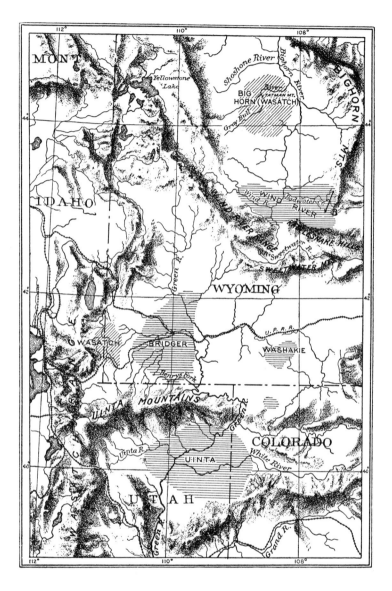

FIGURE 3.11. Location of some of the important fossiliferous Eocene basins mentioned in the text. (From Osborn 1929; courtesy U.S. Geological Survey.)

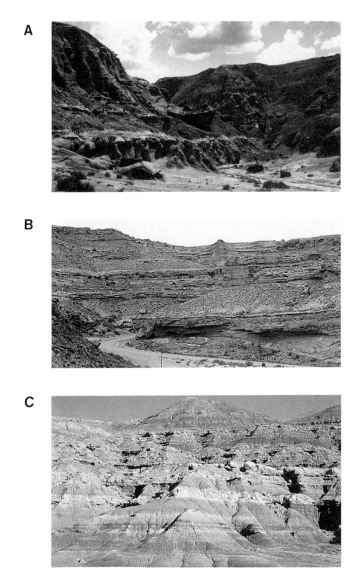

FIGURE 3.12. Outcrops of the classic deposits which are the basis for the middle Eocene in North America. (A) Bridger Formation member B, Bridger Basin, Wyoming. (Photo courtesy R.K. Stucky.) (B) Ledgy sandstones of the lower Uinta Formation, Wagonhound Canyon, Uinta Basin of Utah. (C) Banded red mudstones of the Duchesne River Formation, Uinta Basin of Utah.

interfingers with the Uinta Formation, is the "type" area of the Duchesnean (figure 3.12C).

The Bridger Basin was named for the famous mountain man Jim Bridger, who set up a trading post in the area in 1834. Eventually, the Oregon Trail and the Transcontinental Railroad passed through this region on the way to California. However, the first serious fossil collecting occurred only in 1870, when the pioneering Yale paleontologist Othniel C. Marsh (figure 3.13) led crews to Fort Bridger. The Marsh expedition explored

FIGURE 3.13. Othniel Charles Marsh (back row, center) and Yale College field crew in the Bridger Basin, 1872. Their guns and rough garb were no pose, since the area was still hostile Indian territory at that time. (Courtesy Yale Peabody Museum.)

the purplish buttes and made large collections of middle Eocene mammals. As described by a Yale undergraduate and expedition member, Charles Betts (1871),

A large part of the collection in this region was of the remains of small animals. The fossils were generally found in the buttes, and on account of their minuteness their discovery was attended with much difficulty. Instead of riding along on the sure-footed mule and looking for a gigantic tell-tale vertebra or ribs, it was necessary to literally crawl over the country on hands and knees . . . Often a quarter of a mile of the most inviting country would be carefully gone over with no result, and then again someone would chance upon a butte which seemed almost made of fossils. When two or three found such a prize at nearly the same time, lines would be drawn around each claim with as much care as when valuable mineral land is located; for it must be remembered that each man had full credit for all his discoveries, and the thought of having one's name attached to some rare specimen in the Yale Museum led to sharp competition.

After their success in the Fort Bridger area in August 1870, the Yale expedition moved south and tried to cross the Uinta Mountains. They were soon forced to abandon their wagons in the narrow walls of Flaming Gorge and climb up onto the high table lands near Brown's Hole in the present border region of Utah, Colorado, and Wyoming (now part of Dinosaur National Monument). When they came down on the other side, Betts wrote,

A grand scene burst upon us. Fifteen hundred feet below us lay the beds of another great Tertiary lake. We stood upon the brink of a vast basin so desolate, wild, and broken, so lifeless and silent, that it seemed like the ruins of a world. . . . The intermediate space was ragged, with ridges and bluffs of every conceivable form, and rivulets that flowed from yawning canyons in the mountainsides stretched threads of green across the waste, between their

99

falling battlements. Yet through the confusion could be seen an order that was eternal. For as, age after age, the ancient lake was filled and choked with layers of mud and sand, so on each crumbling bluff recurred strata of chocolate and greenish clays in unvaried succession [now known as the Uinta Formation], and a bright red ridge that stretched across the foreground could be traced far off, with beds of gray and yellow heaped above it [now called the Duchesne River Formation].

Two years after the original Marsh expedition, other paleontologists tried to get in on the bonanza. Joseph Leidy (figure 3.14A), the founder of American vertebrate paleontology, came

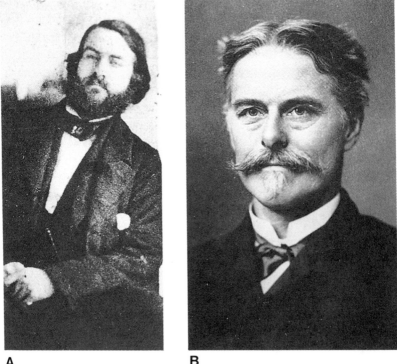

A   **B**

FIGURE 3.14. (A) Joseph Leidy, founder of vertebrate paleontology in North America. (B) Edward Drinker Cope, one of the most brilliant paleontologists of the nineteenth century, and Marsh's bitter rival.

100

to Fort Bridger on his only trip out west. After a lifetime of describing fossils that others had collected from the Big Badlands and many other areas, he got his first chance to see bonehunters' heaven. Leidy (1873) wrote, "No scene ever impressed the writer more strongly than the view of these badlands . . . while walking through the mazes of canyons, it requires but little stretch of the imagination to think oneself in the streets of some vast ruined and deserted city."

The brilliant paleontologist Edward Drinker Cope (figure 3.14B) soon moved in on Marsh's turf. In August 1872 he began collecting in the Bridger Basin, even hiring away some of the locals who knew Marsh's localities. Eventually Cope worked to the east in the Washakie Basin (figure 3.11), which has both Bridgerian and Uintan mammals. Cope and Marsh had long been rivals over naming and describing new species, but the Bridger Basin and its mammals raised their rivalry to out-and-out war. The largest and most spectacular mammals of the middle Eocene were the bizarre uintatheres, huge elephant-size beasts with three pairs of knobby horns on the top of their heads and huge canine tusks (figures 3.1, 3.15). Because of their rivalry, Marsh and Cope rushed to describe their new finds even before they left the field. As they were in remote parts of the country, with limited access to civilization, they had to leave camp to send news east by way of telegraph. In those days, it was common practice to publish a short note of only a few paragraphs naming a new animal, so that you could get credit for being its discoverer and namer. Today, such slapdash methods are frowned upon, but they were common in 1872—especially when trying to beat a rival to press.

Leidy was the first to publish a description of *Uintatherium robustum* (this is now the correct name for most of the specimens). He sent a short note east, dated August 1, 1872. On August 17 Cope sent a telegram from the Black Buttes in the Washakie Basin of Wyoming describing a similar animal. The telegram was badly garbled, however, when it was published two days later; his intended name for the beast, *Loxolophodon*, was misspelled *Lefalaphodon*. The next day another notice was published for Cope that had actually been sent before the telegram of August 17. This later notice named the same beast

101

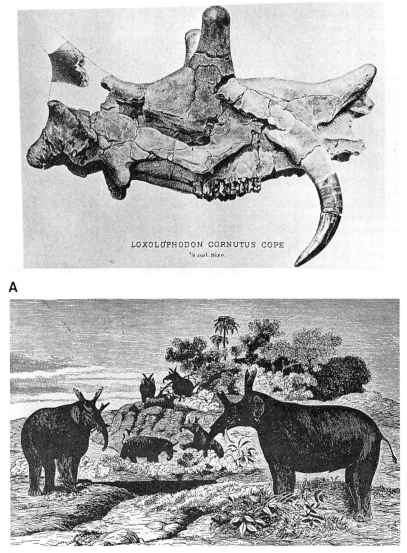

A

B

FIGURE 3.15. (A) A lavish plate from "Cope's Bible" of the skull of the uintathere *Eobasileus* (called *Loxolophodon* by Cope) from the Washakie Basin of Wyoming. (Cope 1884). (B) Cope's elephant-like reconstruction of uintatheres, complete with trunks and large ears. For a more modern reconstruction, see figure 3.1. (From *Pennsylvania Monthly*, August 1973.)

Eobasileus cornutus. Today, this is the valid name for the largest of the uintatheres. On August 22 Cope corrected the garbled name back to *Loxolophodon*, but the name proved unavailable since he had already recklessly used it on another animal years earlier. Meanwhile, Marsh sent a note on August 20 naming his uintathere specimens *Dinoceras* and *Tinoceras* (they are now considered the same genus as Leidy's *Uintatherium*). All three collectors were aware of the others nearby and disputed their rivals' right to collect in "his" fossil field. Soon, this bitter rivalry drove Leidy into retirement from vertebrate paleontology as a field no longer fit for gentlemen.

When Cope and Marsh returned east and began to publish longer descriptions, each became convinced that they both had the same animal—and that only his name for it was correct. Actually, Cope had an *Eobasileus*, and Marsh had a specimen of Leidy's *Uintatherium*, but they considered the differences slight or due to their rival's mistakes. In 1873 Cope compounded his mistakes by suggesting that uintatheres were related to elephants; he even put elephant ears on them in his reconstructions (figure 3.15B). Marsh disputed this, and instead placed them in their own order, the Dinocerata (a name still used today, even though his *Dinoceras* proved invalid). Between August 1872 and June 1873, Cope and Marsh each published 16 articles on uintatheres, each ignoring his rival's names, and both ignoring Leidy's work. As a result, uintathere names reached a state of chaos, with multiple names for the same species. Marsh grew so bitter at Cope's actions that he lashed out in print:

Cope has endeavored to secure priority by sharp practice, and failed. For this kind of sharp practice in science, Prof. Cope is almost as well known as he is for the number and magnitude of his blunders. . . . Prof. Cope's errors will continue to invite correction, but these, like his blunders, are hydra-headed, and life is really too short to spend valuable time in such an ungracious task, especially as in the present case Prof. Cope has not even returned thanks for the correction of nearly half a hundred errors . . . he repeats his statements, as though the

103

Uintatherium were a Rosinante, and the ninth command-
ment a windmill (Marsh 1873)

Eventually, the uintathere wars died down as the rivals
moved into conflicts over the naming of other beasts. Fourteen
years later in 1884, Marsh finally published his uintathere
research, a 237-page tome entitled *Dinocerata: A Monograph
of an Extinct Order of Gigantic Mammals*, with giant folio
pages and lavish plates (figure 3.16). Meanwhile, Cope was
losing ground politically. In the 1870s he had served under
Ferdinand Hayden on the U.S. Geological and Geographical
Survey of the Territories, and from his collections made on
those surveys, he had written a thousand-page, 134-plate
monograph, now known as "Cope's Bible" (figure 3.15A). In
1879 the Hayden Survey was merged with several other government surveys to form the present U.S. Geological Survey. The
first directors were Clarence King and John Wesley Powell,
both good friends of Marsh, and Cope found himself out in the
cold.

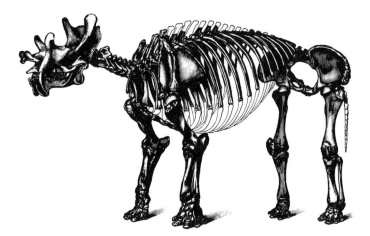

FIGURE 3.16. One of the plates from Marsh's *Dinocerata* monograph,
showing the skeleton of the uintathere *"Tinoceras ingens"* (now
referred to *Eobasileus*). (From Marsh 1886.)

On December 16, 1889, Cope was ordered to turn his collections over to the Smithsonian, even though he had made most of them from private expenditures; they were not paid for by government surveys. Cope was so outraged that he called a reporter, William Hosea Ballou of *The New York Herald*, and filled his ear with grievances against Marsh and his cronies, King and Powell. He charged that they were "partners in incompetence, ignorance, and plagiarism," and that the Survey was a "gigantic politico-scientific monopoly next in importance to Tammany Hall." He leveled charges of every kind against Marsh—that he committed scientific blunders, that he pocketed the salaries of his employees, and that most of his work (especially the *Dinocerata* monograph) was actually the work of assistants.

Marsh defended himself by taking the train to Philadelphia and visiting Cope's bosses, the president and trustees of the University of Pennsylvania. He consoled them about "the shame that has befallen you," suggesting that "poor Cope" had cracked up and that Marsh would help locate "a more substantial scientist" to replace him. In the January 19, 1890 issue of the *Herald*, Marsh replied to Cope's accusations, charging that Cope had stolen Marsh's specimens, and had spied on Marsh's work during a visit to Yale, and had tried to publish it later. Ballou continued to play the feud out for several more columns, quoting and misquoting a number of paleontologists about the scientific competence and personal character of the two rivals. Finally, this particular battle died down, leaving both Cope and Marsh with egg on their faces. Cope retained his position and his fossils, as did Marsh and Powell.

Eventually, though, public scandals did hurt Marsh. When the budget of the U.S. Geological Survey came before a House committee in 1892, fundamentalist congressman Hilary Herbert of Alabama discovered Marsh's recently published monograph entitled *Odontornithes* (on toothed birds from the Cretaceous seas of Kansas). Waving it on the House floor, Herbert shouted, "Birds with teeth! That's where your hard-earned money goes, folks—on some professor's silly birds with teeth." In terms similar to the recent science-bashing of William Proxmire and John Dingell, Herbert stampeded Congress into cutting off funds from such "godless" activities as monographs about impossibilities such as birds with teeth and other creatures not mentioned in the Bible. Powell was forced to send Marsh a telegram: "Appropriations cut off. Please send your resignation at once."

By this point both Cope and Marsh were broken men, and the profession was soon claimed by a new generation. Cope continued to teach at the University of Pennsylvania for five more years, visiting the Dakota Badlands in 1892 and 1893. He died on a cot in his study on April 12, 1897, amidst all his unfinished projects and unpublished specimens. Marsh had spent all of Uncle George Peabody's legacy on his expeditions and lavish publications, so he was forced to live on a modest salary from Yale in a brownstone near the Peabody Museum. In 1896 he published his greatest work, *The Dinosaurs of North America*. Early in 1899, he caught pneumonia; he died on March 18, with less than $100 to his name.

Although the Cope-Marsh feud created a lot of bad feeling, it left an amazing legacy of important specimens that document the middle Eocene in North America. The largest beasts, the bizarre uintatheres, reached the peak of their evolution in the Bridgerian and Uintan. Next in glamor and attention were the brontotheres, or titanotheres, an extinct family of perissodactyls (figure 3.17F). In the early Eocene, early brontotheres like *Lambdotherium* closely resembled the cat-size horse *Protorohippus*, but middle Eocene brontotheres were cow-size (figure 3.17F), and developed blunt paired horns on their snouts.

Although brontotheres and uintatheres were the most spectacular mammals of the middle Eocene, others were more abundant. Horses had evolved into the slightly larger and more advanced *Orohippus* in the Bridgerian and then *Epihippus* in the Uintan (figures 3.1, 3.17B). Primitive tapirs were also common, particularly in the Uintan. The earliest relative of rhinoceroses, known as *Hyrachyus*, was the size of a large dog, but it had almost none of the features (such as horns) we now associate with rhinos (figure 3.17C). Nevertheless, *Hyrachyus* was ubiquitous in the Bridgerian, spreading over all the northern continents.

The most striking event of the middle Eocene, however, was the great radiation of even-toed hoofed mammals. From the

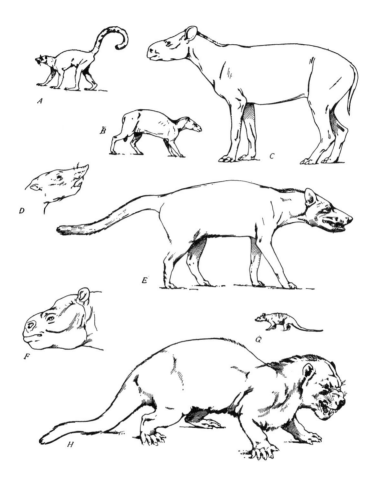

FIGURE 3.17. Reconstructions of Bridgerian mammals. They include (A) the lemur-like primate *Notharctus*; (B) the horse *Orohippus*; (C) the primitive rhinocerotoid *Hyrachyus*; (D) the tillodont *Tillotherium*; (E) the dog-like creodont *Dromocyon*; (F) the primitive brontothere *Palaeosyops*; (G) the anteater-like *Metacheiromys*; (H) the oxyaenid creodont *Patriofelis*. (From Osborn 1909).

107

rabbit-like *Diacodexis* of the early Eocene (figure 1.6D), artio-dactyls diversified into a great variety of families during the Bridgerian and especially the Uintan. Some have living descendants (like the camels, which appeared in the Uintan); some became extinct but have living analogues (several groups were very pig-like or deer-like); still others have no living analogues.

Archaic mammals that had characterized the Paleocene and early Eocene were still around in the middle Eocene, but most were declining. No multituberculates (figure 1.8B) are known from the Bridgerian, and only two genera of these archaic squirrel-like mammals still survived in the Uintan. Bizarre mammals like the tillodonts and taeniodonts (figures 1.7, 3.17D) were extremely rare, and the pantodonts were gone completely from North America by the beginning of the Bridgerian. Among the archaic hoofed mammals, only the dachshund-like hyopsodonts (figure 1.6A) and a relict phenacodont (figure 1.6B) survived into the Bridgerian. The bear-like hoofed mesonychids, source of the whales in the early Eocene, were already scarce by the middle Eocene. The archaic oxyaenid creodonts (figure 3.17H) were also declining, and the 7-foot-tall carnivorous "terror crane" *Diatryma* is last recorded in the Bridgerian (figure 1.5).

The most diverse group of all, however, was the primates (figure 3.17A). Both the adapid and omomyid primates, which arrived in the early Eocene (figure 1.8A), diversified into a great variety of lemur-like forms in the Bridgerian. Only a few genera of rodents were present in the early Eocene but by the middle Eocene, there were dozens of genera and many more species. There was also a great variety of primitive insectivorous mammals living in the underbrush. Several different types of anteater-like animals (the metacheiromyids and epoicotheres) suggest that there were abundant termite and ant colonies to feed upon (figure 3.17G).

Mammals were not the only common vertebrates. The Bridger Formation is famous for its abundant crocodile (figure 3.18) and soft-shelled turtle fossils, indicating that it was deposited in swampy conditions. Both types of reptiles were still present in the Uintan, but they were no longer as diverse or abundant (Hutchison 1982). Middle Eocene river deposits are typically full of many types of fish fossils, especially gar scales.

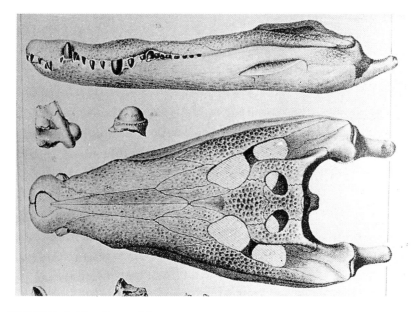

FIGURE 3.18. Crocodiles were common in the Bridgerian. This illustration is from "Cope's Bible" (Cope 1884).

The famous Green River lake shales entomb an enormous variety of fish types, including skates, sturgeons, garfish, bowfins, mooneyes, herrings, suckers, catfish, trout, and perch (Grande 1980). These fossil fishes are so abundant that they are commercially mined; they can be found in rock shops around the country (figures 3.8, 3.19). The Green River shales also preserve a great variety of other lake dwellers, including frogs, lizards, boas, soft-shelled turtles, and crocodiles, plus animals that fell into the lake and were buried, including a perfectly preserved bat and a number of birds. In the arid Uintan deposits in the San Diego region, land tortoises are more common than aquatic reptiles.

Overall, these vertebrate faunas are consistent with the paleobotanical evidence of warm tropical forests in the Bridgerian in most parts of North America. The great abundance of arboreal animals (especially primates) and browsing herbivorous mammals (especially horses, brontotheres, primitive artiodactyls,

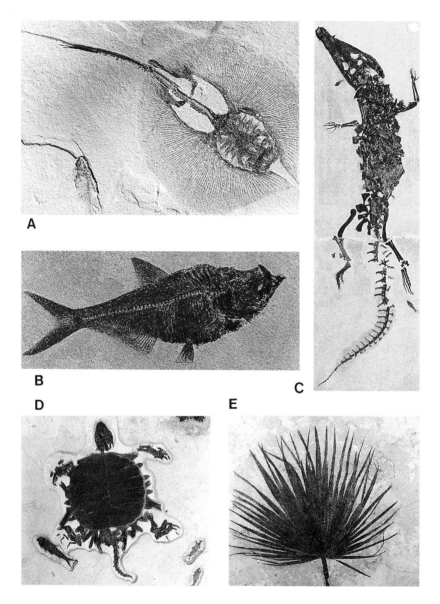

FIGURE 3.19. The middle Eocene Green River Formation produces an array of fossils of extraordinary preservation, trapped in the finely-laminated lake shales. They include (A) a freshwater ray; (B) the bony fish *Diplomystus*; (C) crocodiles, (D) pond turtles, and (E) palm fronds. (Photos C, D, and E courtesy L. Grande.)

110

and archaic hoofed mammals) suggest a forest canopy with abundant leafy vegetation in the lower stories. The striking difference between middle Eocene and earlier faunas, however, is that many of the hoofed mammals (especially brontotheres and uintatheres) are larger in body size than they were in the Paleocene or early Eocene. This suggests that the forest floor was not as densely vegetated, but slightly more open than the early Eocene jungles. The abundance of crocodiles and soft-shelled turtles also indicates an abundandace of fresh water. Despite the floral evidence of slight cooling, Bridgerian faunas are as diverse as early Wasatchian faunas (Stucky 1990). This may be partly the result of the greater diversity of habitats as the uniform tropical forests of the early Eocene broke up into a mixture of paratropical and seasonally dry forests. By the Uintan, this ecological provincialism had greatly increased, and faunas from California were strikingly different from those in the Rocky Mountains or west Texas (Lillegraven 1979).

Although total diversity in the Uintan was as high as in the Bridgerian (Stucky 1990), Uintan faunas show even more evidence of climatic change. Fishes and aquatic reptiles were scarcer, as would be expected with a drying trend that ended the great Green River lake system. Eventually, evaporite minerals were deposited as these lakes dried up. This parallels the drying trend demonstrated in the San Diego Uintan sections (Frederiksen 1991; Peterson and Abbott 1979). As we have already seen, the vegetation was a subtropical woodland, with open areas of grasses, herbs, and shrubby legumes (MacGinitie 1969). Many of the tropical plants that so dominated the early Eocene were replaced by deciduous plants tolerant of a winter dry season.

Given these conditions, it is not surprising that the Uintan shows an even greater scarcity of arboreal mammals; only a few species of primates remained. As the thick undergrowth that sheltered archaic beasts disappeared, so did the last of the tillodonts, pantodonts, phenacodonts, and other archaic jungle herbivores, along with archaic predators, the oxyaenid creodonts. Taking their place was a great diversity of perissodactyls, especially brontotheres and archaic tapirs. By the Uintan, rhinocerotoids had diversified into three great families:

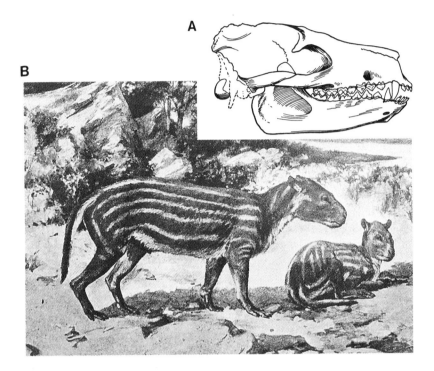

FIGURE 3.20. Oreodonts were primitive artiodactyls, distantly related to camels, but with a great variety of body forms. They were the most common fossil mammals in North America in the middle Eocene and Oligocene, and persisted until the late Miocene. (A) Skull of the Uintan oreodont *Protoreodon*. (B) Restoration of the common Orellan oredont *Merycoidodon*, the commonest fossil from the "Turtle-*Oreodon* beds" of the Big Badlands. (From Scott 1913.)

the aquatic, hippo- or tapir-like amynodonts, the long-legged hyracodonts, and the true rhinoceroses (family Rhinocerotidae). All three groups were probably immigrants from Asia during the Uintan (Prothero et al. 1989).

There were even more types of artiodactyls. These included primitive dichobunids (which were relicts much like early Eocene *Diacodexis*), pig-like achaenodonts, and entelodonts. Many artiodactyls—including the first primitive camels and several extinct groups that are hard to describe in terms of anything living today—had higher-crowned teeth bearing crescent-shaped, leaf-cutting ridges. Some of these animals were similar in size and shape to tiny antelopes. One of the most common Uintan artiodactyls was *Protoreodon*, earliest member of a group known as oreodonts (figure 3.20A). Although they may be distantly related to camels, oreodonts had the proportions and size of small sheep, with extremely primitive skulls, limbs, and other skeletal parts. Oreodonts were North American natives that were restricted to their home continent throughout their long evolution in the Oligocene and Miocene. By the Oligocene, the sheep-size oreodont *Merycoidodon* (figure 3.20B) had become the most abundant herbivore—so common that the lower Oligocene beds of the Big Badlands were once called the "*Oreodon* beds." Oreodonts eventually evolved into pig-like and tapir-like forms in the early Miocene before becoming extinct in the late Miocene.

Clearly, the more open vegetation of the Uintan was favorable to a variety of browsers and mixed-feeding herbivores, including some that were relatively long-limbed and quick, able to dodge in and out of vegetation, and others (brontotheres, uintatheres) that were protected by large body size. The undergrowth also supported a variety of insectivorous mammals, and an explosive radiation of rodents was rapidly taking over the seed- and nut-eating herbivorous niche. In the Uintan, rabbits migrated to North America from Asia, further diversifying the small mammal faunas.

Most of the Uintan predators were primitive carnivorans (including the first members of the dog family, Canidae), no bigger than raccoons. Advanced creodonts known as hyaenodonts occupied a hyaena-like niche (figure 3.21A). Both carnivorans and creodonts had specializations for different types of meat-eating. Some were cat-like, others more bear-like or wolf-like; many had long limbs for pursuing their prey in open country. Many of the oxyaenid creodonts, mesonychids, and earliest carnivorans were, however, generalized omnivores and scavengers: slow-moving and dependent on ambush for hunting.

At the end of the Uintan, North American land faunas suffered their most significant crisis since the end of the Cretaceous. An astounding 80 percent of the genera present in the

Uintan had become extinct by the end of the Duchesnean, and overall diversity plummeted (Savage and Russell 1983; Stucky 1990). Ten families of mammals, mostly archaic groups left over from the Paleocene or early Eocene, disappeared from North America. These included the taeniodonts and uintatheres, the last of the archaic hoofed mammals (hyopsodonts), the last of the North American primates (adapids and paromomyids) and colugos ("flying lemurs"), and several archaic insectivorous groups. These extinctions at the end of the middle Eocene coincided with a severe cooling and drying trend, and this must have caused a crisis in the subtropical Uintan forests that had sheltered these animals. Temperatures and floras did recover slightly in the late Eocene (discussed in the next chapter), but the splendid forest-dwelling community of mammals did not.

Unfortunately, the fossil record for the Duchesne River Formation is very sparse, leading some authors to question whether a distinct Duchesnean land mammal "age" is justified (Emry 1981; Wilson 1978, 1984, 1986). Such sparse collections pose problems for deciding how much of the post-Uintan diversity crash and mass extinction is due to poor sampling. Fortunately, a number of other regions (from Oregon and California to Saskatchewan to South Dakota to Texas) produce Duchesnean mammals and give us a much more complete sample of this 3-million-year interval (Kelly 1990; Lucas 1992). Stucky (1992) found that total Duchesnean diversity is comparable to many other land mammal ages with much bigger samples, so the post-Uintan decline appears to be real.

Although Duchesnean faunas are relatively sparse, several important families first appear at this time. These include the true rhinoceroses (Family Rhinocerotidae), which were first represented in North America by *Teletaceras* from the Clarno Formation of Oregon (Hanson 1989). Only the size of a German shepherd dog, *Teletaceras* had front teeth diagnostic of the family: the chisel-like upper incisors occluding with lower tusks. *Toxotherium*, a controversial animal that has been variously called an amynodont rhino, a tapiroid, or a chalicothere, was probably an immigrant lophiodont escaped from Europe (Prothero et al. 1986; Hooker 1989; Schoch 1989a). Also

114

FIGURE 3.21. Two types of mammals typical of the "White River Chronofauna" which immigrated to North America in the Duchesnean. (A) The scavenging creodont *Hyaenodon*, last member of this archaic group. In the background are the tiny deer-like *Leptomeryx*. (B) The giant pig-like entelodont *Archaeotherium*, with huge bony protuberances on its face. (From Scott 1913.)

FIGURE 3.22. The "White River Chronofauna" also contained these Duchesnean immigrants. (A) The running rhino *Hyracodon*, about the size of a Great Dane. (B) The primitive long-tailed, clawed oreodont *Agriochoerus*. (From Scott 1913.)

prominent were the pig-like entelodonts, first known in North America from *Brachyhyops*. It was soon replaced by the big White River entelodont, *Archaeotherium*, which persisted well into the Oligocene, filling the role of scavenger and rooter (figure 3.21B). The advanced creodont *Hyaenodon* (figure 3.21A) was an important newcomer in the Duchesnean, as were the earliest members of the "beardog" carnivorans, the family Amphicyonidae. A number of new rodent genera appeared as well, along with primitive beaver relatives, the Eutypomyidae.

Other Duchesnean genera were probably descended from indigenous Uintan ancestors, although they came to dominate the White River Chronofauna. These included the peculiar insectivore *Leptictis*, the squirrel-like ischyromyid rodents, the Great Dane–sized running rhino *Hyracodon* (figure 3.22A), the three-toed horse *Mesohippus*, the long-tailed and clawed oreodont *Agriochoerus* (figure 3.22B), and the tiny deer-like artiodactyls *Leptomeryx* and *Hypertragulus*. By the end of the Duchesnean (middle late Eocene), a few more of the characteristic archaic Uintan groups had vanished, including the last of the lemur-like primates, the carnivorous hoofed mesonychids, and the very last of the archaic ungulates: the dachshund-like hyopsodonts. These disappearances indicate that the dense forests of the Bridgerian and Uintan were gone for good.

The Asian Connnection

By the Bridgerian, North America's connection with Europe across the North Atlantic had been severed, and immigration between these two regions ceased. Both areas developed their own endemic faunas. However, later on in the Uintan there is evidence of increased migration between North America and Asia. Mongolia, China, and Burma all produce remarkable middle and late Eocene mammal faunas. As in the early Eocene, Asia was home to a great number of relict archaic animals, including the eurymylids (source of rodents and rabbits), tillodonts, didymoconids, and the last of the oxyaenid creodonts: a huge scavenger called *Sarkastodon* (figure 3.23A). The Mongolian uintathere *Gobiatherium* (figure 3.23B) had a huge bulbous snout instead of the multiple horns and tusks of American uintatheres. There were still relict pantodonts in Asia

117

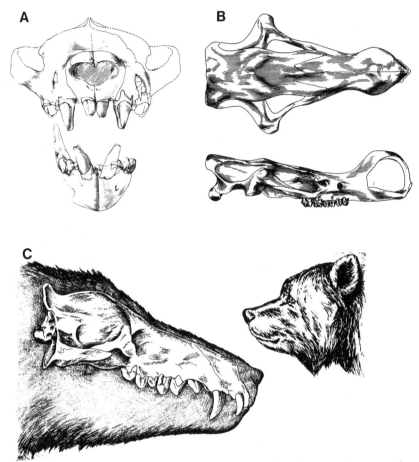

FIGURE 3.23. Middle Eocene deposits of China and Mongolia produced some spectacular beasts, most of which were extinct by the late Eocene. They included (A) the huge, bear-like bone-crushing creodont *Sarkastodon*. (From Granger 1938). (B) The bulbous-snouted, hornless uintathere *Gobiatherium*. Compare with figure 3.1 and 3.15A. (From Osborn and Granger 1932.) (C) The skull of the gigantic Mongolian mesonychid, *Andrewsarchus* (left) compared with a modern Kodiak bear, the largest living carnivore (right). No skeleton of *Andrewsarchus* is known, but it is conventionally given a bear-like body. Considering that mesonychids gave rise to whales (see figure 1.14), it may have had a much more aquatic shape. (From Fenton and Fenton 1958; by permission of Doubleday, Inc.)

long after they had died out elsewhere, and mesonychids were still abundant. The most spectacular of these was the gigantic *Andrewsarchus*, known only from a skull almost three feet long and two feet wide—more than twice the size of the largest bear that ever lived (figure 3.23C). Because the skull is wolf-like or bear-like, the animal is often reconstructed as if it were a gigantic bear. If its skeleton were bear-like, it probably would have been twelve feet long and more than six feet high at the shoulder. However, since mesonychids were ancestral to whales, it is possible that the skeleton of *Andrewsarchus* was much more like a primitive whale, which would make its size more believable. Unfortunately for this hypothesis, the skull was found in terrestrial deposits.

Among the more controversial members of the middle Eocene of eastern Asia are fossils of primates from the Pondaung fauna of Burma (Colbert 1937; Ba et al. 1979; Ciochon et al. 1985). Two primates, *Pondaungia* and *Amphipithecus*, have been called the earliest relatives of monkeys and apes, although they may be the same age as very primitive anthropoids recently discovered in the Eocene of Africa. There is still much argument as to whether monkeys and apes originated in Africa or in southeast Asia during the middle-late Eocene.

The most striking feature of this time, however, is the large number of advanced mammals present in both the Uintan of North America and the Asian middle Eocene (especially Irdin Manha and Shara Murun in Mongolia). Both continents shared a great diversity of brontotheres; some genera were common to both regions. Both continents also had a number of archaic tapirs, although Asia supported a greater diversity. Hyracodontid (figure 3.22A) and amynodont rhinos seem to have roamed between the continents with ease, as did a variety of early artiodactyls, especially the pig-like entelodonts (figure 3.21B), the deer-like leptomerycids, and the aquatic anthracotheres. According to Savage and Russell (1983), the mesonychid *Harpagolestes*, the entelodonts, several genera of hyracodont and amynodont rhinos, the claw-bearing perissodactyls known as chalicotheres, and the rabbits were all clearly emigrants from Asia to North America in the Uintan.

119

THE COOLING BEGINS

The transition from the middle to the late Eocene in Asia is confused by recent adjustments in correlation. Because the Uintan has been recorrelated with the middle Eocene and the Chadronian with the late Eocene, many of the Asian localities usually called "early Oligocene" (Li and Ting 1983; Russell and Zhai 1987; Wang 1992) are probably late Eocene (see Berggren and Prothero 1992 for detailed discussion). A comparison of the Irdin Manha and Shara Murun faunas of the middle Eocene (Li and Ting 1983; Russell and Zhai 1987) with the late Eocene (mislabeled "early Oligocene") faunas of China (Wang 1992, table 27.1) shows striking changes at the end of the middle Eocene in Asia. Rabbits and rodents continued to be common, along with the last remnants of their primitive relatives, the anagalids. Primates were still around, although greatly reduced in numbers. Archaic mesonychids and tapirs still persisted in small numbers, but most other typically Eocene groups had disappeared: pantodonts, tillodonts, oxyaenid creodonts, archaic hoofed mammals, and the peculiar didymoconids. As in other continents, the dominant trend was diversification of perissodactyls—particularly brontotheres, chalicotheres, and hyracodont and amynodont rhinocerotoids. Artiodactyls (particularly the pig-like entelodonts, and deer-like animals, and the ubiquitous anthracotheres) were also more common after the end of the middle Eocene.

In summary, the middle-late Eocene transition in Asia, as in North America, was a severe extinction event that eliminated many arboreal and archaic mammals typical of the early and middle Eocene forests. Although an exact tabulation of valid species has not yet been done, this extinction crisis is comparable to the contemporaneous reductions in North America at the end of the Uintan. However, much more detailed stratigraphic analysis and dating is needed to determine how many of these extinctions occur right at the middle/late Eocene boundary, and how long the crisis lasted.

European Eocene Archipelagos

The rise in sea level during the middle Eocene isolated western Europe. Not only was the North Atlantic corridor severed, but the northward expansion of warm, tropical waters divided

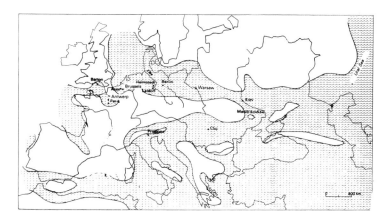

FIGURE 3.24. The European archipelago during the middle and late Eocene, showing the major marine transgression and isolated islands of endemic mammals. (Modified from Pomerol 1976.)

Europe into an archipelago of islands in a shallow tropical sea (figure 3.24). The paleobotanical record from the middle Eocene of Europe shows a cooling trend similar to that recorded by North American floras. But in Europe the cooling and drying were ameliorated by the warm tropical seas and the island humidity. Margaret Collinson notes only slight changes in the vegetation in southern England at the beginning of the middle Eocene (Collinson and Hooker 1987). There was a slight decline in diversity among subtropical and tropical plants (including the ubiquitous paw-paw trees and moonseed and icacina vines) and an increase in freshwater plants. *Nypa* palms (figure 1.10A) were rare by the early middle Eocene and had vanished by the end of the middle Eocene. The abundant mangrove swamps of the early Eocene also disappeared in the middle Eocene. As the floral change of the early-middle Eocene was rather subtle in North America, so too in Europe.

The cooling at the middle-late Eocene transition is much more apparent in the English floras (Collinson and Hooker

121

FIGURE 3.25. Reconstruction of palaeotheres from the middle Eocene of the Paris Basin, with a tapir-like lophiodont in the right distance. Crocodiles (foreground) are also common in these deposits. (Parley 1837).

1987). Late Eocene floras were dominated by reed marshes with abundant cattails, pondweeds, and leather ferns, occasionally broken by shrubby patches of deciduous cypresses (Taxodiaceae), walnuts, *Papaver* poppies, buttercups, raspberries, and blackberries. Almost all the tropical elements characteristic of the early Eocene had disappeared by the late Eocene.

Boulter (1984) examined the Paleogene floras of the Hampshire Basin of southern England and estimated paleotemperatures based on the tolerances of modern analogues and descendants. He shows a peak in mean annual temperature of about 23°C (73°F) in the early Eocene, followed by a sharp decline to about 16°C (61°C) at the early-middle Eocene transition. This decline was comparable to the temperature changes reported by Wolfe (1978) for the same interval. The end of the middle Eocene in Europe was marked by a mean annual temperature of only 13°C (55°F). In the late Eocene, there was another warming to about 21°C (70°F), similar to the late Eocene warming reported by Wolfe (1978).

As a consequence of geographic isolation through most of the middle and late Eocene, European mammals evolved separately from the rest of the world and developed their own endemic groups (Collinson and Hooker 1987; Hooker 1992). Instead of the abundant true horses, brontotheres, tapirs, and rhinocerotoids found in North America and Asia, European perissodactyls consisted mainly of a horse-like group known as palaeotheres (figure 3.25), descended from early Eocene *Hyracotherium* (which Hooker 1989 argues was not a horse). In addition, there was a group of tapir-like perissodactyls known as lophiodonts. Both of these groups were unique to Europe during the middle and late Eocene, although a few lophiodonts spread into North America in the late Eocene. Both palaeotheres and lophiodonts had teeth with leaf-crushing cross-crests, much like modern tapirs, so they were clearly forest folivores.

The most spectacular specimens of these beasts come from deposits at Messel in Germany (see Schaal and Ziegler 1992 for outstanding photographs of these amazing fossils). Located about 20 miles (30 km) southwest of Frankfurt, the site was once a large open-pit oil-shale mine (figure 3.26A). When the site was threatened by a proposed sanitary landfill in 1975, an international scientific program saved it and the study of its classic fossils intensified. Messel is unique—even in the priceless class of exceptional fossil deposits. Animals are found as complete articulated skeletons in death poses with every bone in its proper position. Even more remarkable is the fact that the outlines of bodies are also preserved as a black film on the shale surface. In a fossil frog from Messel it is possible to distinguish eyes, liver, and veins; in fossil birds, the detailed structure of feathers is preserved; in fossil bats the complete wing membrane is still visible. Most remarkable, however, is that the stomach contents of animals are still

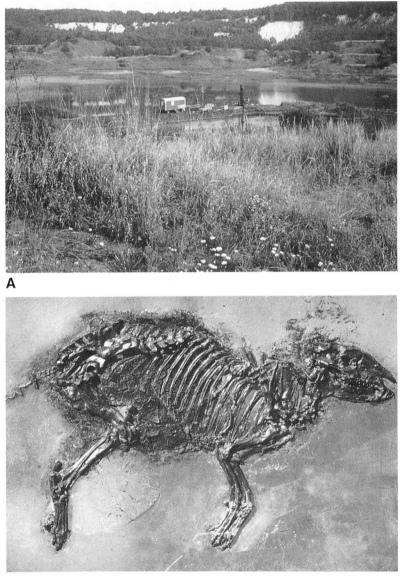

FIGURE 3.26. (A) The famous Messel locality, near Frankfurt, Germany. (B) Complete articulated skeleton of *Propalaeotherium*, a horse-like mammal from Messel. (C). *Messelobunodon*, a primitive artiodactyl. (Photos courtesy J. Franzen.)

C

preserved. The shales record, for example, that one kind of bat dined exclusively on butterflies.

Among these striking Messel animals is a complete specimen of the horse-like perissodactyl *Propalaeotherium* (figure 3.26B). Preserved in a complete articulated death pose, the body outline and the delicate, seldom-preserved bones in the throat and ear region are even visible. The most valuable aspect of the fossil, however, is the preservation quality of its stomach contents: a mass of clearly identifiable leaves and fruit seeds show conclusively what palaeotheres ate. Another specimen turned up with very tiny bones in the pelvic region. When "baby teeth" were also found, it confirmed that these were bones of an unborn embryo still in the mother.

How did these remarkable fossils become so exquisitely preserved? The Messel deposits are shales formed in a very small (a few square miles in area) but deep lake in a fault valley. The bottom of the lake did not have any circulation or overturn, so it was stagnant and without oxygen, thereby preventing scavenging fish and microorganisms from living there. In the warm, subtropical climate, frequent algal blooms even further depleted the oxygen at the bottom and increased its depth stratification. When animals or plants fell in or carcasses floated in, they sank to the bottom where they were buried before they could decay. The only scavengers were bacteria adapted to oxygen-poor environments. The mysterious black outlines have been studied and were found to be made entirely of the mineralized remains of these bacteria.

In addition to *Propalaeotherium*, Messel also yields complete skeletons of an anteater, a pangolin, a variety of insectivores, and a primitive dichobunid artiodactyl, *Messelobunodon* (figure 3.26C). Indeed, middle Eocene Europe had an even greater variety of unique artiodactyls in addition to its endemic perissodactyls. Along with dichobunids (holdovers from the early Eocene), there were a number of families (cebochoerids, choeropotamids, mixtotherids, dacrytherids, haplobunodonts, xiphodonts, amphimerycids, cainotheres, anoplotheres, anthracotheres) that are known only to European specialists. Most of these animals lived only in the Eocene of Europe, and did not spread to any other continent (except for aquatic anthracotheres, which were global). As in the great North American radiation, some European artiodactyls were very pig-like, and others resembled tiny antelopes; xiphodonts might possibly be related to camels. Most of these animals, however, had no relatives or descendants outside of Europe.

Although the middle Eocene fauna of Europe was highly endemic, there were many ecological similarities with other regions. Lemur-like primates were abundant in the forest canopy, although they were adapids and omomyids unique to Europe. Hyaenodont creodonts were more common than true carnivorans, although some paleontologists suggest that the earliest members of the bear and mongoose families were already present. The most common animals were the rodents. They were the most endemic group of all, dominated by families that never occurred outside Europe and that mostly became extinct in the early Oligocene.

Hooker (in Collinson and Hooker 1987) records only small changes in the mammalian fauna at the beginning of the middle Eocene. The palaeothere *Hyracotherium* with low-crowned teeth was replaced by perissodactyls more suited for leaf-eating, such as *Propalaeotherium*, several species of lophiodonts, and the rhinocerotoid *Hyrachyus*. Primitive rodents of the subfamilies Microparamyinae were replaced by the more squirrel-like Pseudosciuridae. Hartenberger (1986, 1987) shows that archaic beasts like multituberculates, tillodonts, pantodonts, and colugos all disappeared from Europe at the end of the early Eocene, leaving a fauna dominated by perissodactyls, artiodactyls, rodents, and primates. As in the Wasatchian-Bridgerian transition of North America, the early-middle Eocene transition in Europe was subtle.

The end of the middle Eocene in Europe differed from its counterpart in North America or Asia in that it was not a major extinction event (Hooker 1992). The dominant herbivores in the late Eocene were tapir-size leaf-eaters like *Palaeotherium* and *Plagiolophus*. The browsing Theridomyidae replaced the fruit-eating Pseudosciuridae among the rodents. Both of these replacement groups had teeth with better-developed crests for browsing coarser leaves. Overall, Europe experienced a drastic reduction in arboreal mammals (particularly primates) and reduction in insectivorous mammals, but an increase in large ground mammals and browsing herbivores. The last of the European archaic hoofed mammals finally disappeared at the end of the middle Eocene. These changes parallel some of those seen at the end of the Uintan in North America, but they were not nearly as severe.

FIGURE 4.1. The spiny planktonic foraminiferan *Hantkenina alabamensis*, whose last appearance is used to recognize the Eocene/ Oligocene boundary. (Photo courtesy G. Keller.)

Terminal Eocene Event?

Post hoc, ergo propter hoc
"After this, therefore because of this"

—A FAMOUS LOGICAL FALLACY

*The great tragedy of science—the slaying of a
beautiful hypothesis by an ugly fact.*

—THOMAS H. HUXLEY, *BIOGENESIS AND ABIOGENESIS* (1894)

*Facts are simple and facts are straight
Facts are early and facts are late
Facts all come with points of view
Facts don't do what I want them to*

—TALKING HEADS, *CROSSEYED AND PAINLESS* (1980)

Thanks to decades of Hollywood stereotypes, most people visualize scientists as socially maladjusted types with white lab coats, thick glasses, and pocket protectors, standing before a bench full of bubbling beakers and sparking electrical apparatuses. In almost all these films, the stereotypical scientists recite obligatory lines about how they must objectively seek the facts, or how they cannot allow their personal feelings to get in the way of objective truth. If the scientist is given any human dimension at all, he (movie scientists are rarely female) is either a modern Dr. Frankenstein unleashing some horrible lab creation on the world, or, at best, totally naive about the effects of his work on humans.

Judging from recent surveys, these misconceptions about science and scientists still prevail in the public mind. To some extent, this is the fault of scientists and science educators as well. Science courses rarely discuss the human dimension of science, emphasizing instead the understanding and memorization of facts. Yet science is very much a human endeavor, and scientists cannot eliminate their human sides from their work any more than nonscientists can.

As social animals who must survive in a particular professional community, scientists also demonstrate behaviors that would astonish most other people. Indeed, a whole subfield of the sociology of science has arisen in recent years to study these activities. Scientific meetings can degenerate into shouting matches and name-calling, although the preferred method of attack is to demolish one's opponent with a witty riposte. As David Hull documents in his book *Science as a Process* (1988), the maneuvering and backstabbing that goes on with grants and publications in a particularly contentious field of science can be truly amazing.

Scientific Fads and Mass Extinctions

One of the striking sociological features of science is a well-known politicial and social activity called "jumping on the bandwagon"—latching on to a trendy idea when it is rolling to get a free ride. Politicians do it when a particular candidate, party, or ideology has irresistible momentum to put them in power. Scientists do it when a particular theory or idea gives one a better chance of getting a grant, publishing a paper, or landing a book contract. Mass extinction research is a particularly vivid example of the bandwagon effect.

The notion that a major change took place between the Eocene and Oligocene is not new. As early as 1909, the Swiss paleontologist Hans Stehlin noticed a striking difference between Eocene and Oligocene mammals of Europe, which he called "La Grande Coupure" ("the great break"). Henry Fairfield Osborn (1910) observed the same phenomenon in his book *The Age of Mammals in Europe, Asia, and North America.* Many publications about the marine realm, such as A. Morley Davies's (1934) *Tertiary Faunas* and Cifelli's (1969) classic study of planktonic foraminiferans came to similar conclusions. We have known about the great differences between the Eocene and the Oligocene for almost a century. In more recent

130

years, papers by Black and Dawson (1966), Gregory (1971), Lillegraven (1972), Wilson (1972), and a detailed review by Webb (1977) pointed out the striking turnover in fossil mammals without provoking heated controversy or intense research.

Jack Wolfe (1971) was struck by the North American floral evidence of a dramatic cooling which he called the "Oligocene deterioration." In his 1978 paper, however, changes in the time scale led him to relabel this cooling the "Terminal Eocene Event." From the paleobotanical data, it was clear that something significant had happened around the end of the Eocene, but uncertainties in correlation and time scales hampered any further investigations. There was also increasing evidence from the Deep Sea Drilling Project (summarized by Kennett 1977 and Frakes 1979) that significant oceanographic changes occurred at the Eocene-Oligocene transition. However, few could make the connection between North American land plants and Antarctic marine cooling, because problems with correlation were very difficult.

Then the stakes changed radically. In 1980 Luis and Walter Alvarez, Frank Asaro, and Helen Michel published a paper which suggested that extinctions at the end of the Cretaceous (including the death of the dinosaurs) were caused by an asteroid impact. They found unusual concentrations of iridium, a metal in the platinum group, in the K/T boundary clay at Gubbio, Italy. Iridium is rare in crustal rocks but abundant in the mantle and in meteorites. The impact hypothesis triggered an explosion of papers—pro and con—over the next decade. Mass extinction research expanded dramatically as everyone tried to get in on the act, and the research soon divided into schools of thought, with the century-old battle between "catastrophists" and "gradualists" re-emerging. Tempers grew hot, and even Nobel Prize–winning physicist Luis Alvarez resorted to unseemly name-calling. As David Raup wrote in his 1991 book, *Extinction, Bad Genes or Bad Luck?*,

Mass extinction is box office, a darling of the popular press, the subject of cover stories and television documentaries, many books, even a rock song. . . . At the end of 1989, the Associated Press designated mass

131

extinction one of the "Top 10 Scientific Advances of the Past Decade." Everybody has weighed in, from the *Economist* to *National Geographic*.

After the 1980 Alvarez et al. study, the bandwagon was accelerated by another key paper. In 1984 David Raup and Jack Sepkoski claimed that mass extinctions recurred approximately every 26 million years (figure 4.2). Since they could not envision any earthly causes with a 26-million year periodicity, they suggested that there might be some sort of extraterrestrial event that occurred every 26 million years. As recounted in Raup's 1985 book, *The Nemesis Affair*, he and Sepkoski sent preprints of their manuscript to a number of astronomers *before* it was published (and thus before it had a chance to be evaluated by other scientists to decide whether it was worth taking seriously). Predictably, astronomers jumped on the bandwagon, proposing several hypotheses that might explain a 26-million-year periodicity. These ranged from periodic comet showers (Davis et al. 1984), to the oscillation of the solar system through the galactic plane (Rampino and Stothers 1984; Schwartz and James 1984), to an unknown Planet X (Whitmire and Jackson 1985) and even an undetected companion star of the sun (Whitmire and Jackson 1984). Whitmire and Jackson even went so far as to name this star Nemesis although it had not yet been proven to exist. The response to the periodicity preprint was so rapid that four of these five articles were published in sequence in the same issue of *Nature* just months after the original Raup and Sepkoski paper appeared in the *Proceedings of the National Academy of Sciences*. As Keith Thomson (1988) put it, "Inevitably, with most subjects there is also a silly season, usually of unpredictable duration and of an intensity correlated with the state of acceptance of the new idea . . . [including] proposal of ideas even more far-out than the original one."

Unfortunately for the pro-impact stampede, several ugly little facts killed their beautiful hypotheses. Repeated efforts to find Planet X or Nemesis, or to tie the extinction cycle to comet showers and movements through the galactic plane, have all fizzled (Shoemaker and Wolfe 1986; Tremaine 1986; Sepkoski

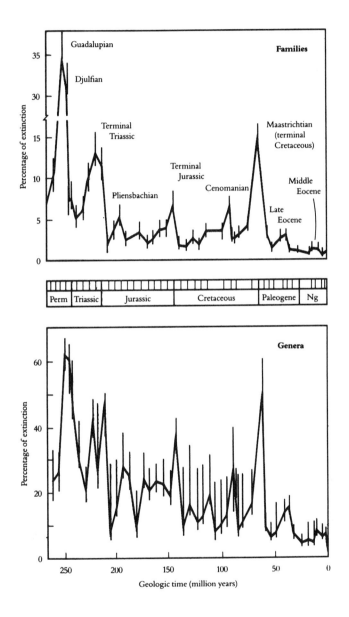

FIGURE 4.2. Percentage of extinction in marine families and genera since the Permian. Middle Eocene "extinction peak" is shown. (Modified from Raup and Sepkoski 1986.)

133

1989). Even more problematic is the alleged 26-million-year extinction cycle. A number of statisticians and paleontologists (Kitchell and Pena 1984; Kitchell and Easterbrook 1986; Hoffman and Ghiold 1986; Harper 1987; Stigler and Wagner 1987; Noma and Glass 1987; Quinn 1987) have argued that Raup and Sepkoski's statistical methods do not establish a true periodicity, or that the "periodicity" could be due to random noise in the data. Sepkoski (1989) tried to answer these arguments, but there is still much doubt among paleontologists. Another criticism focuses on the nature of the data. Some paleontologists, familiar with the problems of naming and defining fossil families and genera, claimed that Raup and Sepkoski's conclusions were invalid because they relied on a data base full of bad taxonomy (Hoffman 1985, 1989; Patterson and Smith 1987, Smith and Patterson 1988). This accusation is particularly damning, because if the data base is not real, then all the arguments about statistics are moot. For example, Smith and Patterson (1988) showed that when mistakes in identification or dating of fossils were eliminated from the data base of echinoderms and fishes used by Raup and Sepkoski, the "periodicity" disappeared.

Critics have also questioned whether all the extinction peaks are real. For example, the middle Miocene "extinction peak" at 13 million years is based on only a few species of molluscs, and it does not show up in the excellent record of Miocene land mammals (Webb 1977). The early Jurassic peak was barely above background noise levels, and Sepkoski (1989) has since abandoned all pretext of a mid-Jurassic extinction event. Some alleged peaks (such as the late Triassic, the mid-Jurassic, the early Cretaceous, and the Pliocene) fall well outside the predicted time interval (Sepkoski 1989). If only half these peaks now appear to be real and on schedule, and there are long gaps with no extinction at the predicted 26-million-year interval, what happens to periodicity?

Another problem in establishing periodicity is the variability of time scales (Hallam 1984; Hoffman, 1985, 1988, 1989; Stanley 1987). As indicated in chapter 2, some parts of the geologic time scale have fluctuated wildly in the last decade. Raup and Sepkoski (1984, 1986) predicted that the Eocene-Oligocene

134

extinction event should have occurred at 39 million years ago, yet the date of the Eocene/Oligocene boundary has fluctuated between 38 and 32 million years ago. The only evidence of a possible impact occurs in strata around 35.4 to 36.0 million years old. The boundary itself is around 34 million years ago, and the greatest extinctions occurred at 37 and 33 million years. Shoemaker and Wolfe (1986) found only three extinction horizons dated accurately enough for inclusion in periodicity studies, and these proved insufficient to establish periodicity. Ironically, one of the three was the Eocene/Oligocene boundary.

Another criticism of the periodicity proponents is the practice of lumping all the extinction data into stages with durations of several million years. This procedure treats all extinctions in a given interval as if they were concentrated at the end, even if they were spread out through the entire stage. As a consequence, the data base bunches all extinctions at the stage boundaries, regardless of when they occurred. Clearly, there are problems with treating the Eocene-Oligocene extinctions as a single event exactly 26 million years after the Cretaceous extinctions.

In the case of the Eocene-Oligocene transition, the stampede for impacts and periodicity threatened to overwhelm any sensible attempts to examine the data in detail. Once the iridium was found in upper Eocene rocks and the nearest extinction was attributed to the Raup and Sepkoski periodicity, some scientists treated the case as proven without further discussion. At the 24th International Geological Congress in Washington, D.C. in 1989, paleontologist Digby McLaren argued that *all* mass extinctions were caused by impacts, *whether or not there was evidence of impact in the fossil record!* David Raup (1991) has written that all extinctions (even normal "background" extinctions) might be caused by impacts. With this kind of fuzzy logic and inattention to the constraints of evidence, why bother gathering data anymore? Extinctions occurred, and we know that impacts have also occurred—therefore impacts caused extinctions. *Voila!*

The claim of impact extremists can be falsified by examining impacts that are identified in strata that have no associated extinctions. The Montagnais impact structure, located off the shelf of Nova Scotia, has been dated at about 50 million years ago (late early Eocene), which corresponds to no extinctions of significance (Bottomley and York 1988; Aubry et al. 1990; Jansa et al. 1990). A similar conclusion has been reached about the Ries crater, the result of an extraterrestrial object that hit Germany in the Miocene (Heissig 1985).

But even where there is coincidence or correlation (very shaky in the case of the late Eocene) between mass extinctions and impacts, this still does not establish causation. Philosophers of logic refer to this kind of error as the *post hoc* fallacy, from the Latin aphorism *Post hoc, ergo propter hoc* ("After this, therefore because of this"). In Edmund Rostand's play *Chantecler*, the rooster thought that because the sun rose each morning after he crowed, he caused it to rise. Statisticians are very careful not to overstate the meaning of correlation. Even though certain phenomena appear to be correlated, it is an entirely different matter to prove that one caused the other. In many cases, the two events may coincide because they have a common underlying cause, not because one caused the other. If the number of coincidences is relatively small, then there is a high probability that they occurred by chance.

Some scientists, dismayed by the inflated claims of impact advocates, decided to study the Eocene-Oligocene transition in detail, and to closely scrutinize the evidence of impacts. Raup and Sepkoski had treated all the extinctions of the entire middle and late Eocene as if they occurred precisely at the end of the Eocene, even though they spanned at least seven million years. With the convenient "Terminal Eocene Event" label already in the literature, neo-catastrophists assumed that the Eocene-Oligocene transition was a single catastrophe like the terminal Cretaceous extinction. Indeed, the sloppy but convenient habit of lumping everything into a "Terminal Eocene Event" focused attention on this boundary in particular. The title of a 1985 symposium on "Terminal Eocene Events" (Pomerol and Premoli Silva 1986) reinforced this unfortunate tendency to focus on the boundary, even though the text of the conference proceedings clearly shows that the transition was prolonged over millions of years.

How Many Impacts? Did They Really Matter?

After the 1980 discovery of the K/T iridium anomaly, scientists looked for evidence of impacts everywhere, and soon they focused on the "Terminal Eocene Event." In 1982 several scientists (Alvarez et al. 1982; Asaro et al. 1982; Ganapathy 1982; Glass et al. 1982) reported slightly increased concentrations of iridium in late Eocene limestones in the Caribbean. In some of the same sections, glassy spherules known as tektites (figure 4.3) were found (Donnelly and Chao 1973; Glass and Zwart 1979; Glass et al. 1979, 1982, 1985). Tektites and microtektites (which are smaller than 1000 microns) are pieces of crustal rock melted and thrown into the atmosphere by the impact of an asteroid (Taylor 1973). The correlation of this layer with the Eocene/Oligocene boundary or any major extinction horizon was not clear, but at the time, it didn't matter. All that counted was that there was *some* association of extraterrestrial debris and extinctions, and the case was closed in the minds of those who advocated impact causes of all mass extinctions.

As these tektites were examined further, however, complications arose. The first problem was that the impacts were the wrong age. The inital reports of microtektites and iridium from a deep-sea core in the Caribbean, followed by further reports in Barbados and the Pacific (figure 4.4), all placed the horizons within the late Eocene (planktonic foraminiferan Zone P15, or magnetic Chron C16N), well before the Eocene/Oligocene boundary. Direct dating of microtektites (Glass and Crosbie 1982; Glass et al. 1986) eventually produced $^{40}Ar/^{39}Ar$ dates of 35.4 ± 0.6 million years. At the time, this seemed close to the estimated age of the boundary, even though it was two planktonic foraminiferan zones too early. Now that we have better estimates of the age of boundary, it is clear that the tektites are about 1.5 million years too early to have had any effect.

As more scientists examined additional deep-sea cores, another problem arose: there were *too many* tektite layers! Following reports by Billy Glass and colleagues of a single layer (Glass 1974; Glass and Zwart 1977, 1979; Glass et al. 1973, 1979), Gerta Keller and her co-workers found evidence of three layers (Keller et al. 1983). An intense debate ensued (Glass 1984; Keller et al. 1984; Glass 1986; Keller 1986), with

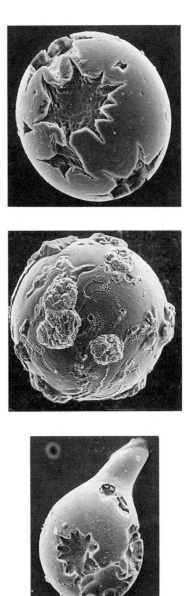

FIGURE 4.3. Examples of glassy impact droplets known as microtektites from the late Eocene. Each is about 100 microns in diameter. (Photos courtesy G. Keller.)

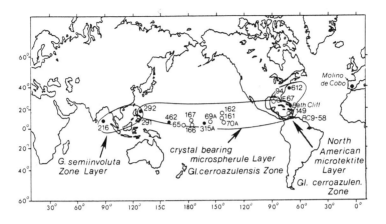

FIGURE 4.4. Map of major tektite strewn fields of late Eocene age. (From Keller et al. 1987.)

Glass et al. (1985) finally admitting to two layers in the Caribbean, and Keller et al. (1987) and D'Hondt et al. (1987) insisting there was a third distinct layer in the Pacific. Hazel (1989) used graphic correlation methods on multiple sections of the late Eocene and came up with *six* events! Not surprisingly, Hazel's conclusions were also contested (Glass 1990, and reply by Hazel 1990). Byerly et al. (1990) presented evidence that two of Hazel's layers were artifacts of modern contamination, but Hazel persisted in maintaing that there were six. Montanari (1990) found only two iridium anomalies superposed in the late Eocene section in Gubbio, Italy; these data provided an additional check of the controversial sections in the Western Hemisphere. Only one of these iridium anomalies, moreover, contained microtektites, and both of them occurred in the middle of the late Eocene, well before the Eocene/Oligocene boundary.

After so much controversy, an independent third-party investigation was needed. Miller et al. (1991a) examined the evidence anew, and concluded that four discrete horizons could be substantiated. The oldest layer occurs in planktonic

139

foraminiferan Zone P15 and in the top of magnetic Chron C16N1–2 (about 36.0 million years ago). It contains both microtektites and iridium, and appears both in the Pacific and in the western Atlantic near New Jersey. The second layer, evidenced by tektites off the shore of New Jersey, was a few thousand years younger than the first—although it, too, occurs within Zone P15 and Chron C16N1–2. The impact sites of both these events were probably in the western Atlantic.

A good candidate for the site of one of these impacts was recently found on the edge of the Atlantic continental shelf, about 100 km east of New Jersey (Poag et al. 1992). Drilling in the region uncovered evidence of a blanket of impact-ejected debris over 50 m thick, buried 1000 m below the sea floor in water almost 200 m deep. Although the crater's shape and outline are still vaguely defined, it appears to be about 15 km in diameter, with a raised rim more than 200 m higher than the adjacent late Eocene beds. Based on the known scaling of craters, Poag and colleagues estimated that it was formed by an impacting object about 1 km in diameter.

The effects of this impact off New Jersey can be traced far into the Atlantic coastal region, where is has been correlated to a well-known upper Eocene subsurface deposit known as the Exmore boulder bed. This boulder bed was discovered by drilling in Chesapeake Bay; it contains boulders over 2 m in diameter and shock-metamorphosed quartz grains characteristic of impacts. The boulder bed suggests a high-energy wave, similar to the tsunamis produced by earthquakes. In some places, the boulder bed is over 60 m thick, and it may extend more that 15,000 square kilometers beneath the Atlantic Coastal Plain from New Jersey to North Carolina (Poag et al. 1992).

The third horizon described by Miller et al. (1991a) is a crystal-bearing microspherule layer with iridium. It occurs in planktonic foraminiferan Zone P16 and magnetic Chron C15R, and is dated at 35.5 ± 0.3 million years (Obradovich et al. 1989). This third layer has been found throughout the Pacific, in the Caribbean, and in marine strata in Italy. The fourth and final discrete layer was correlated with the well-known layer of large tektites in the southeastern United States known as the

North American strewn field. It is also within planktonic foraminiferan Zone P16 and Chron C15R, and it is dated at 35.4 ± 0.6 million years (Glass et al. 1986). The North American strewn field is found primarily in the Caribbean and Atlantic, and also in marine sediments in Italy.

Thus, there appear to have been two closely spaced impact events about 36 million years ago (in Zone P15, Chron C16N1–2), followed 600,000 years later by two closely spaced events about 35.4 millon years ago (in Zone P16, Chron C15R). No crater or impact site has been located for either of these final two impacts, and given the wide dispersal of debris from the Pacific to Italy, one can only guess where the impact site might be. Because the Eocene/Oligocene boundary occurs about 34 million years ago (end of Zone P17 and late in Chron C13R), all of these impacts were too early to have affected the boundary events.

One final test of the impact-extinction hypothesis: do we actually *observe* any extinctions in sediments with impact debris? Although five radiolarian species do become extinct near the level of the 35.4 million-year-old tektites in Barbados (Maurrasse and Glass 1976; Glass and Zwart 1977; Sanfilippo et al. 1985), there are no extinctions in the foraminiferans (Keller, in Hut et al. 1987). Five radiolarian species (out of dozens) is not a particularly impressive number by the standards of the extinctions we see in the rest of the Eocene and Oligocene. Four foraminiferan species appear to have become extinct in the upper part of Zone P15, and in association with microtektites, but the extinctions and microtektites may be fortuitously combined by a dissolution horizon, which would artificially truncate the stratigraphic ranges of the microfossils and concentrate microtektites at the same level (figure 4.5). These strata do show fluctuations in abundances and size changes in certain planktonic foraminiferans (Keller 1986; Keller et al. 1987; Keller, in Hut et al. 1987; MacLeod 1990) associated with impact debris, but this is far from the global catastrophe that the impact hypothesis predicts. Hansen (in Hut et al. 1987) suggested that the late Eocene molluscan extinctions in the U.S. Gulf Coast (which occur in the middle of Zone P15) might be correlated with one of the earlier impact

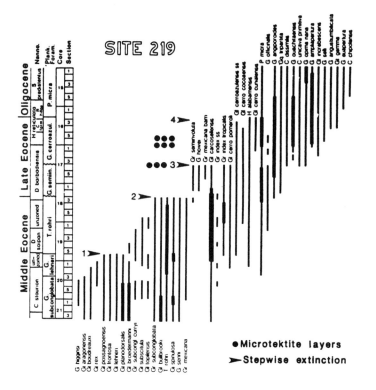

FIGURE 4.5. Middle-late Eocene extinctions in planktonic foramini-ferans, and levels of microtektites, from DSDP Site 219 in the Indian Ocean. Event 3 may be due to a condensed, dissolution horizon (wavy lines). Two microtektite layers above this correspond to no significant extinctions (from Hut et al. 1987, by permission of *Nature*.)

horizons, but later (1992) Hansen admitted that the correlation is approximate, and there is no direct evidence to associate them. More importantly, there is no evidence of extinction in the other tektite horizons of the late Eocene, and no evidence of extinction in organisms other than a few microplankton and possibly some molluscs.

142

We must conclude that the late Eocene impacts were both too early and too late to have been a significant cause of Eocene-Oligocene extinctions, and the few biotic changes that may be correlated with them are decidedly minor. Yet the literature even before the impact frenzy treated the Eocene/Oligocene boundary as the biggest biotic catastrophe of the entire Cenozoic, complete with its own label: the "Terminal Eocene Event." What actually happened at the boundary?

Boundary Disputes

Chapter 2 examined how Lyell's method of defining the Eocene led to many arguments about which stages were actually part of the Eocene and which were Oligocene. The primary problem was Lyell's failure to designate type sections or to use stratigraphic ranges of fossils to mark boundaries. Instead, he conceived of the Eocene as a "moment" on the clock face—a moment, moreover, that had no specific duration, type sections, or boundaries. The problems due to Lyell's vagueness were compounded by the fact that von Beyrich's type Oligocene, which were based on strata in Belgium and Germany, does not overlie Lyell's "type" Eocene beds of the Paris Basin or the Italian Apennines. None of the "typical" areas, therefore, can be used to draw the boundary.

Disputes over the stratotypes of the late Paleogene stages (see Berggren 1971) could not resolve the problem. Type areas of these stages do not lie in the same basins, and most of their fossils are shallow marine molluscs that can not be correlated across Europe, let alone around the world. Subsequent research (see Hardenbol and Berggren 1978; Berggren et al. 1985) has shown that there were time gaps between the classic European stages as originally defined, leaving some Eocene and Oligocene intervals without a name.

It fell to the micropaleontologists to solve this dilemma (Berggren 1971; Berggren et al. 1985). Deep-marine sequences, full of rapidly evolving, widespread foraminiferans and coccoliths, were far more complete and fossiliferous than terrestrial or shallow-marine strata. Deep-marine strata offered a much better chance of defining temporal boundaries with worldwide expression and without significant time gaps between

143

units. To minimize fruitless wrangling over terminology, stratigraphers try to find the most representative deep-marine section
uplifted and exposed on land for a particular boundary and
then, by vote of committee, they designate a formal type
section. This method has already been applied to a number of
stratigraphic boundaries (see Prothero 1990 for a review). In
each case, a variety of suitable "type sections" is suggested,
with consideration of the abundance of key fossils, accessibility
of exposures, and lack of unconformities or structural com-
plexities. Then, over the course of several years, the committee
visits most of the candidate outcrops to complete the documentation before a vote is taken. Derek Ager (1973) recommended
establishing some actual physical marker (metaphorically,
driving a "golden spike") on the exact level of the boundary in
the type section. This would conclusively resolve disputes
about how a particular boundary is defined and recognized,
although there is always going to be argument about how to
correlate it with other areas.

To define the Eocene/Oligocene boundary, stratigraphers
had to ignore Lyell's criteria and examine the various stages
proposed in Europe (Berggren 1971; Hardenbol and Berggren
1978; Cavelier 1979; Berggren et al. 1985). Several named
stages were relevant to the boundary. The Priabonian Stage was
proposed by Munier-Chalmas and de Lapparent in 1893, based
on exposures near Priabona in the Vicenza province of northern Italy. The Priabonian was an alternative to the Paris Basin
"Ludian" (also proposed by Munier-Chalmas and de Lapparent
in 1893). The Ludian was dropped by de Lapparent in 1906,
because it could not be recognized in other regions. The type
Priabonian contained planktonic foraminiferan Zones P15 to the
lower part of P17, and nannofossil Zones NP18 to lower NP21.
The Rupelian Stage was proposed by Dumont in 1849 for
exposures of the "argile de Rupelmonde" ("argile de Boom," or
"Boom Clay") at Boom (Anvers), Belgium. It spanned the upper
part of planktonic foraminiferan Zones P18 to the upper part of
P19, and nannofossil Zone NP23. Thus, neither the Priabonian
nor the Rupelian stratotypes had fossils representing upper
planktonic foraminiferan Zones P17 or lower P18, or upper

nannofossil Zones NP21 or NP22, which span the Eocene/ Oligocene boundary.

A third stage name commonly used was the "Lattorfian," proposed by Meyer-Eymar in 1893 for the Latdorf Sands in Saxony, Germany. These glauconitic sands and their molluscs were thought to represent the lowermost Oligocene, earlier than the type Rupelian in Belgium. However, detailed analysis of molluscs and nannofossils from the Lattorfian showed that it spanned much of the late middle Eocene to earliest Oligocene (Berggren et al. 1985). The Lattorfian was clearly unsuitable as a stage, moreover, because the type section is in an abandoned lignite mine in eastern Germany that has been inaccessible for over eighty years (Ritzkowski 1981). The term "Tongrian," proposed by Dumont in 1839 for sequences in Belgium, expanded in usage until it was almost synonymous with von Beyrich's later term "Oligocene." Eventually, the meaning of Tongrian was restricted to the Sables de Neerrepen (now considered late Eocene) and to fluvial-marine sediments containing the famous Hoogbutsel mammal locality (long thought to be early Oligocene). Since it spanned the Eocene/Oligocene boundary and was not nearly as fossiliferous or complete as the type Priabonian or Rupelian, the Tongrian too has been abandoned (Cavelier 1979). Today most stratigraphers recognize the Priabonian as the only stage name for the late Eocene, and Rupelian alone for the early Oligocene.

With such incomplete shallow marine stage stratotypes for reference, the deep sea produced much more complete sequences of marine microfossils, and these became the standard. A number of different biostratigraphic markers could have been used to pin down the boundary. Several genera of planktonic foraminiferans became extinct through this interval, as did a number of nannofossils. Martini and Ritzkowski (1968) recommended defining the boundary on the last appearance of rosette-shaped discoasters (*Discoaster barbadiensis* and *D. saipanensis*), which were characteristic late Eocene coccoliths. However, the last occurrence of these coccoliths appears to have been controlled by temperature and was therefore time-transgressive (Cavelier 1972; Berggren et al. 1985).

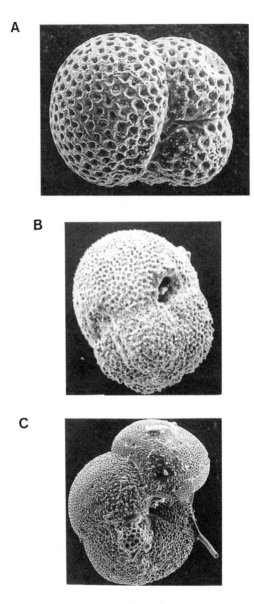

FIGURE 4.6. Representative late Eocene planktonic foraminiferans. (A) *Subbotina lineapertura*; (B) *Globigerinatheka semiinvoluta*, zonal indicator of the early late Eocene; (C) *Turborotalia cerroazulensis*, zonal indicator of the late late Eocene. Each specimen is about 300 microns in diameter. (Photos courtesy G. Keller.)

In the early 1980s, International Geological Correlation Project 174 was convened to determine criteria for recognizing the Eocene/Oligocene boundary. The committee met in 1985 and published their results in 1986 (Pomerol and Premoli Silva 1986). The preferred candidates for the boundary marker were the last appearances of several characteristic late Eocene foraminiferans (figure 4.6), including the spiny forms *Hantkenina* (fig. 4.1) and *Cribrohantkenina, Globigerinatheka, Turborotalia cerroazulensis*, and the large *Pseudohastigerina micra* and *P. danvillensis*. Also considered were the first appearance of the early Oligocene foraminiferan *"Globigerina" tapuriensis* and the increase in abundance of *Turborotalia ampliapertura*. Most of these events were clustered at the boundary between planktonic foraminiferan Zones P17 and P18 (Nocchi et al. 1986). The overturn of so many species at the P17-P18 zonal boundary suggested that this was the best candidate for the Eocene/Oligocene boundary.

Along with biostratigraphic studies, several exposed outcrops bearing these fossils were proposed for the type section of the Eocene/Oligocene boundary. Two sections in the Italian Apennines (the famous Gubbio sections, where the K/T iridium was first found, and another at Massignano) produced good sequences of microfossils with magnetic stratigraphy and datable ash layers. The other main contenders were Bath Cliff in Barbados and the Betic Cordillera in southern Spain, but these were rejected because they were less complete, or lacked the datable minerals or magnetics found in Italian sections.

Documentation of the Italian sections was spearheaded by Isabella Premoli Silva and her colleagues, who produced a volume which showed that all the necessary elements were available in the Italian sections (Premoli Silva, Coccioni, and Montanari 1988). They recommended the section at Massignano (figure 4.7), on the Adriatic Coast of Italy near Ancona, as the best candidate for a stratotype (Nocchi et al. 1988; Odin and Montanari 1988). Exposed in an abandoned quarry, it will probably always be accessible to scientists. Its micropaleontology, magnetostratigraphy, and lithostratigraphy have been carefully documented, and it has also produced some dated

147

FIGURE 4.7. Stratotype section òf Eocene/Oligocene boundary at Massignano, near Ancona, Italy. The boundary is now designated the last appearance of *Hantkenina*, at the 19-meter level in this section, on the uppermost left side of the central saddle. (Photo courtesy A. Montanari.)

ash layers. At the 19-meter mark on the section, the Zone P17 foraminiferans disappeared from the limestones. As the boundary must be based on a single biostratigraphic event (several would be confusing), the last appearance of the spiny planktonic foraminiferan *Hantkenina* (figure 4.1) was recommended as the biostratigraphic datum that marked the end of the Eocene. Even in places where planktonic foraminiferans were unavailable, the magnetic stratigraphy at Massignano and Gubbio showed that *Hantkenina* last appeared in the upper third of magnetic Chron C13R; the Eocene/Oligocene boundary could thus be recognized in nonmarine rocks as well.

At the 24th International Geological Congress in Washington, D.C. in July 1989, the International Subcommission on Paleogene Stratigraphy met and considered the recommendation of Premoli Silva and her colleagues. After some debate, they voted in favor of the Massignano section. The "golden spike" had been "driven" at the 19-meter mark on the Massignano quarry face, and disputes about the boundary seemed to be over.

However, problems have arisen. Brinkhuis (1992) studied dinoflagellates (organic-walled microscopic cysts) from the same Italian limestones. He found that the dinoflagellate that disappeared alongside *Hantkenina* vanishes 15 meters below the top of the type section of the Priabonian. If the Priabonian is the basis for the late Eocene, then the last 15 meters of the stratotype of the late Eocene are early Oligocene! Brinkhuis (1992) pointed out that the top of the type Priabonian section corresponds to a major sea level drop (and as it turns out, to a variety of other events, including cooling and glaciation, Wolfe's floral deterioration, and the "Grande Coupure"), all of which are good indicators of the Oligocene. If these criteria were used instead, the Eocene/Oligocene boundary would fall in magnetic Chron C13N, about a million years younger than the *Hantkenina* datum. Several authors (e.g., McGowran et al. 1992) favor the events within Chron C13N as the choice of boundary, because it is controlled by a global climatic signal and because major unconformities and isotopic signals are far easier to recognize than the last appearance of a single microfossil. As Berggren (in Berggren and Prothero 1992) points out, "realism, and a respect for the history of the debate on chronostratigraphic boundaries would suggest, however, that we have not heard the last on the subject of the Eocene/Oligocene boundary."

If, however, we use the *Hantkenina* datum, what *does* occur at the Eocene/Oligocene boundary? As documented by Corliss et al. (1984) and Snyder et al. (1984), the boundary itself is pretty quiet. In spite of the changes in the planktonic foraminiferans, the benthic foraminiferans show little or no change, nor do the calcareous nannofossils and diatoms. The carbon and oxygen isotopic ratios were fairly stable across the boundary.

TERMINAL EOCENE EVENT?

Prior to the boundary (at the end of the late Eocene, planktonic foraminiferan Zone P17), Boersma et al. (1987) show isotopic and foraminiferan evidence for a general cooling and decrease in oceanic circulation, resulting in a less stratified ocean. Late in Zone P17, there was a decrease in surface water temperatures that was reponsible for the disappearance of rosette-shaped discoasters, followed by the extinction of warm-water foraminiferans such as *Hantkenina* and the other boundary markers. But as the next chapter will show, the biggest global changes happened in the early Oligocene, about a million years after the extinction of *Hantkenina*. (Unless, of course, we redefine the boundary again to correspond with the oxygen isotope shift and all the related changes.) In short, the Eocene/Oligocene boundary (as now defined) was one of the more uneventful times in geologic history.

Why, then, does it have the reputation of a catastrophe? The confusion is due to bad correlation and conflation of several discrete events into a single crisis. The big changes in sea level, ice volume, oxygen isotopes, and Wolfe's floral cooling all occurred in what is now defined as early Oligocene (magnetic Chron C13N, about 33 million years ago). Until recently, most sedimentary records across the boundary interval were very incomplete, or had such slow sedimentation rates that the *Hantkenina* datum was just below the isotope shift. Under these conditions, researchers could not distinguish between the last appearance of Eocene microfossils and the early Oligocene refrigeration event. But recent coring in the deep sea around Antarctica, as well as more careful work on the Italian sections, have shown that these were discrete events. In addition, research on the entire spectrum of changes through the Eocene and Oligocene has shown that most of the extinctions occurred at the end of the middle Eocene (as discussed in the last chapter). In short, the Eocene-Oligocene "event" is actually several events, spread over at least 7 or 8 million years. The Eocene/Oligocene boundary itself marks one of the least significant phases in this transition, signalled primarily by turnover in planktonic foraminiferans.

Late Eocene Rebound: Calm Before the Storm

Detailed studies of the late Eocene show that it was actually quite an interesting interval. According to the latest calibrations, the late Eocene spanned about 3 million years between the middle Eocene extinctions at about 37 million years ago and the beginning of the Oligocene at about 34 million years ago. Most oxygen isotopic records (Shackleton 1986; Corliss and Keigwin 1986; Miller et al. 1987; Miller 1992) show a slight decrease (about 0.5‰) in the oxygen isotope ratio, indicating a warming of about 1°C by the middle of the late Eocene.

In marine organisms, this slight temperature fluctuation doesn't seem to have had much of an effect, compared to the long-term cooling and oceanographic changes. Temperature-sensitive calcareous nannofossils show a slight warming in the late Eocene (Haq and Lohmann 1976), but most other organisms were not so responsive. Instead, marine organisms retained their low diversity levels from the terminal middle Eocene extinctions, before suffering further in the early Oligocene cooling. In planktonic foraminiferan Zone P15 (about 37 to 35 million years ago, or early late Eocene), there were no significant changes in response to this slight warming (Boersma et al. 1987). The record of Gulf Coast molluscs (Hansen 1988, 1992) contains only two late Eocene data points, so it is hard to tell if there was a late Eocene recovery. The signal is dominated by a dramatic decline from the high in diversity just before the end of the middle Eocene. Data from echinoids (McKinney et al. 1992) have even poorer resolution for the late Eocene. Not until the latter part of the late Eocene (planktonic foraminiferan Zones P16–P17, about 34 to 35 million years ago) do we see dramatic oceanographic changes, and these are really the beginning of the Oligocene deterioration, which will be the focus of the next chapter.

The rebound on land was much more dramatic. Based on leaf-margin analysis of North American land plants, Wolfe (1978) suggests a mean annual temperature increase during the late Eocene from 20°C (68°F) to 28°C (82°F) in the U.S. Gulf Coast, and from 12°C (54°F) to 18°C (64°F) in the Gulf of Alaska region (figure 3.7). This 6–8°C (10–14°F) warming was comparable to the recovery during the early middle Eocene. Once again, North American continental climate seems to amplify small changes in global temperature.

In botanical terms, Wolfe (1978) sees the broad-leaved evergreen forests found at the end of the middle Eocene in the Pacific Northwest returning to paratropical rain forests by the middle of the late Eocene. This fluctuation is documented by superposed floras in several localities. For example, in central Oregon the paratropical flora of the middle Eocene lower Clarno Formation (figure 3.10) is overlain by the broad-leaved deciduous floras of the middle Clarno assemblage, which (as discussed in the previous chapter) seems to correspond to the cooling at the end of the middle Eocene. Overlying it, however, is the late Eocene upper Clarno flora, which again contains paratropical plants (Wolfe 1971). The Puget Group of western Washington produces the middle Eocene Steels Crossing flora, which is broadleaved evergreen, followed by broadleaved deciduous floras of the cooler lower Ravenian (middle-late Eocene transition). During the warmer upper Ravenian (late Eocene), broadleaved evergreen floras returned that match those of the upper Clarno (Wolfe 1971). In the Gulf of Alaska, broadleaved evergreen forests of the middle Eocene were succeeded by more cold-tolerant broadleaved deciduous forests near the middle-late Eocene boundary, then returned to warmer broadleaved evergreen forests in the late Eocene (Wolfe 1969, 1971).

The floras of the Rocky Mountains do not show this temperature fluctuation (Wing 1987). However, in areas like the Big Badlands, ancient soil horizons give a vivid picture of the changing landscape. According to Greg Retallack (1981, 1983a, 1983b, 1990, 1992), early late Eocene forests received about 1 m of annual precipitation. By the latest Eocene, these moist forests had changed to a dry woodland with open patches of grasses and herbs. This allowed archaic browsers like the brontotheres to persist, while the more varied, open habitat gave room for a variety of mixed-feeding herbivores, some of which could eat tougher grasses and herbs.

FIGURE 4.8. Low rounded knobs of deeply weathered bentonitic clays characteristic of the late Eocene Chadron Formation in Badlands National Park, South Dakota. The banded siltstones of the early Oligocene Scenic Member of the Brule Formation are visible in the distance.

Brontothere Boneyards

The response of land vertebrates to these climatic changes in the late Eocene is just now being documented. In North America, the late Eocene is represented by the Chadronian (about 34–37 million years ago) land mammal "age" (Emry et al. 1987; Prothero and Swisher 1992). It took its name from the Chadron Formation, the base of the sequence in the White River badlands in South Dakota, Nebraska, and adjacent states (figure 4.8). When first explored, these strata were called the "*Titanotherium* beds" (Meek and Hayden, 1857; Osborn 1910, 1929) because rich "graveyards" of brontotheres have eroded out of the greenish-gray clays that weather to a "popcorn"

153

A

B

FIGURE 4.9. (A) John Bell Hatcher, Marsh's chief collector in the Badlands. He collected hundreds of brontotheres in just a few years. (B) Henry Fairfield Osborn, President of the American Museum of Natural History at the beginning of the twentieth century, and founder of its Department of Vertebrate Paleontology. He is best known for his monographs on brontotheres, rhinos, horses, proboscideans, and other large vertebrates. (Negative no. 109272, courtesy Department of Library Services, American Museum of Natural History.)

surface. The great collector John Bell Hatcher (figure 4.9A) spent the summer of 1866 working out of Chadron, Nebraska. Between early May and October, he collected an astounding 25,000 pounds of brontothere bones for Yale University. The next March he was back in the field, collecting in eastern Wyoming, where he excavated 11 brontothere skulls. When he moved on to the Big Badlands of South Dakota, he collected another 13 skulls, 3 in a single day! Before he moved on to other things, he had collected more than 200 skulls for O. C. Marsh and for Yale (Lanham 1973).

Brontotheres were at the climax of their evolution in the Chadronian, reaching elephantine sizes with huge, blunt paired horns on their snouts. Their skulls were the prime goal of Badlands collectors, and many of the major eastern museums made large collections from the "titanothere graveyards" (actually, ancient river channel bottoms that tended to accumulate the largest and heaviest bones). Turn-of-the-century paleontologists such as O. C. Marsh and Henry Fairfield Osborn (figure 4.9B) were impressed with these big beasts, and their magnificent mounts were the pride of exhibition halls. Marsh and Hatcher never got a chance to publish the definitive monograph on the brontotheres before they both died, but the rich collections of specimens in eastern museums was eventually described by Osborn in a gigantic two-volume, 951-page monograph published in 1929. Like most paleontologists of his time, Osborn was a "splitter." Every new specimen with some slight difference from other specimens was described as a new species. Not until the 1940s and 1950s did paleontologists and systematists begin to realize that natural populations of species can be highly variable. Today, most paleontologists take this variability into account, and "lumping" many highly variable samples into a single species, unless there are good reasons for separating them. They look at the total sample of specimens from a single bed, or single horizon, and try to determine how much is due to natural variation (or sexual differences between males and females), and what differences really distinguish biological entities we call species.

In the case of brontotheres, Osborn (1929) was a hypersplitter; he named new lineages that few of his contemporaries took seriously (Rainger 1991). Because of Osborn's powerful position as President of the American Museum of Natural History and chair of its Department of Vertebrate Paleontology (the center of the paleontological profession at that time), few scientists could afford to argue with him. In many cases, Osborn named new species for skulls that looked different simply because they had been crushed or deformed. In the titanothere monograph, he left a legacy of dozens of genera, split into no less than twelve subfamilies, thereby making brontotheres appear more diverse than any other group of mammals.

155

Osborn was legendary for his aristocratic attitudes, pomposity, and monumental ego, revealed in nearly everything he wrote (Rainger 1991). He was particularly famous for promoting his notion of "aristogenesis." Surveying the gradual enlargement of brontothere nasal horns, he could not see how Darwinian natural selection could act on the tiny nubbins on the snouts of early brontotheres (figure 4.10). According to Osborn, the horn enlargement must have been driven by internal, non-Darwinian forces. When brontotheres died out at the end of the Chadronian (which Osborn thought was Oligocene), it was because they had reached "racial senescence." The horns had evolved out of control, eventually becoming maladaptive.

Osborn thought that his titanothere monograph would be the definitive work on the subject, requiring no further revision. His egoism reached its zenith in a book entitled *Fifty-Two Years of Research, Observation, and Publication* (1930). There Osborn wrote,

In the summer of 1877 I saw in the Bridger Basin of southwest Wyoming my first fossil titanothere, and ten years later—namely, in 1887—I began the series of seventeen papers and memoirs on the titanothere family (Brontotheriidae). This led to my appointment in the year 1900 as vertebrate palaeontologist of the United States Geological Survey and to twenty-nine successive years of as arduous intensive research as has ever been pursued in any central field of science, culminating in my Survey monograph No. 52 . . . of 951 pages, 795 figures, 236 plates, by far the most profound study of a central family history [of animals] that has ever been made; this monograph is entitled *The Titanotheres of Ancient Wyoming, Dakota, and Nebraska* (pp. 157–158)

Although his contemporaries knew the work was seriously flawed, the sheer size and bulk of the two-volume titanothere monograph intimidated paleontologists for over sixty years. Everyone knew that the systematics of brontotheres were a mess, but no one wanted to spend the time that it would take to clean up Osborn's chaos. It would require years of work in the

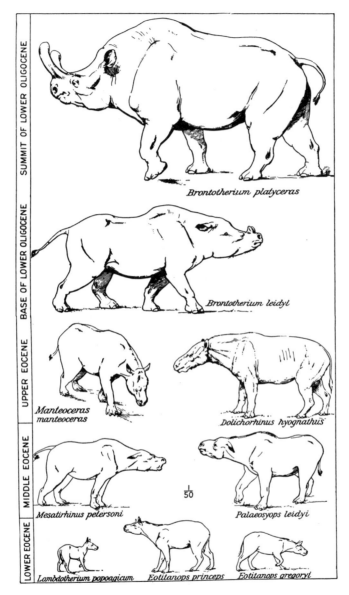

FIGURE 4.10. Osborn's (1929) ideas of brontothere phylogeny, emphasizing his notions of linear "aristogenetic" evolution. Most of the taxonomic names shown here are now invalid. Osborn's "upper Eocene" is now middle Eocene, and "lower Oligocene" is now late Eocene. (Courtesy U.S. Geological Survey.)

157

collections at Yale, the American Museum in New York, the Smithsonian, and other institutions.

Since the 1930s, dozens of brontothere skulls (especially from the Chadronian) had been collected by the Frick Laboratory of the American Museum. All of these skulls had excellent stratigraphic data (unlike the turn of the century "hit and run" collections), so it would be possible to sort out how brontotheres changed through time, as well as examine large samples from the same time horizon to determine population variability. Unfortunately, those collections are still in their plaster jackets and field wrappings, and it will take years of expensive preparator time to make them ready for study. Several eager students who considered brontotheres as a possible Ph.D. thesis soon changed their focus when they ran into this "plaster wall."

After sixty years of neglect by generations of paleontologists, Bryn Mader (1989) made a first stab at cleaning up the Osbornian quagmire. Out of dozens of genera and twelve subfamilies, Mader now recognizes only 18 valid North American genera in only two subfamilies for the entire Eocene. Middle Eocene brontotheres were undergoing rapid evolution and diversification in the Uintan (especially as they grew horns), but that evolution was not nearly so rapid and diverse as portrayed by Osborn. The "straight-line" evolution of brontotheres with increasing horn lengths (still often portrayed in textbook diagrams recycled from Osborn) was a figment of Osborn's imagination (figure 4.10). Although horn length did increase through time, species of different horn lengths lived side by side.

Sorting out which species are valid will be a much bigger task, and will require breaching the "plaster wall" around the Chadronian brontotheres in the Frick Collection. For the moment, Mader (1989) recognizes only three valid genera of Chadronian brontotheres, replacing a chaos of invalid names. *Megacerops* (formerly known as *Brontotherium* and *Titanops*) stood 8 feet (2.5 m) tall at the shoulder, had a skull that was elongated and high behind the eyes, yet held only a small brain, and had a pair of horns growing from the snout. These horns were made of bone and were probably covered with skin, much like the horns of the giraffe. In *Megacerops* each horn was long and flaring, with a circular cross-section. Another

158

Chadronian brontothere, *Menops* (including the subsumed genus *Allops*) had horns with a more triangular cross-section. Mader recognizes only one more valid genus of Chadronian brontothere: *Brontops* (also known as *Diploclonus*), which had short, forward-pointing horns. Almost every name you see in the popular books and reprinted, out-of-date Osbornian diagrams—including such classic names as *Brontotherium*, *Titanotherium*, *Menodus*—are invalid.

The brontothere's world of the late Eocene was inhabited by a mixture of jungle relicts from warmer times (including not only brontotheres, but also multituberculates, primates, archaic tapirs, and even crocodiles and pond turtles), along with many new groups of mammals. The late Eocene was truly a world of transition between the early Eocene rainforests and the colder, drier world of the Oligocene. The most abundant animals in the late Eocene and Oligocene were very different from the assemblage of primates, multituberculates, archaic ungulates, pantodonts, and creodonts that were typical of the Paleocene and early Eocene.

After the great decline in crocodilians and aquatic turtles at the end of the Uintan, reptilian diversity stabilized in the late Eocene (Hutchison 1982, 1992). The Chadronian is distinguished by the high rate of evolutionary turnover, seen especially in the appearance of new mammals. Most of these new groups appear to be immigrants from Asia, accentuating the trend that began in the Uintan and Duchesnean. Many Asian immigrant groups came to dominate North American communities well into the late Oligocene, forming a distinct ecological unit that Emry (1981; Emry et al. 1987) calls the "White River Chronofauna."

The Chadronian was marked by even more immigrants from Asia. Their appearance further adds to the list of modern families that made up the White River Chronofauna, giving North America a much more recognizable aspect (figure 4.11). Chadronian debuts include the first true squirrels (*Protosciurus*, Family Sciuridae), true beavers (*Agnotocastor*), pocket gophers (Geomyidae) and pocket mice (Heteromyidae), the first North American peccaries (pig-like mammals, also known as javelinas, which are still common in the southwestern U.S. and Latin

FIGURE 4.11. Characteristic mammals of the Chadronian-Orellan-Whitneyan-early Arikareean "White River Chronofauna." 1—the pig-like entelodont *Archaeotherium*; 2— the gazelle-like camel *Poebrotherium*; 3—the common oreodont *Merycoidodon*; 4—the long-tailed oreodont *Agriochoerus*; 5—the primitive horse *Mesohippus*; 6—the false sabertooth *Hoplophoneus*; 7—the amphibious anthracothere *Bothriodon*; 8—the creodont *Hyaenodon*; 9—the running rhino *Hyracodon*; 10—the multihorned camel relative known as *Protoceras*; 11—the hornless true rhinoceros *Subhyracodon*. Small dog shown for scale. (From Scott 1913.)

America), the pig-like anthracotheres (figure 4.11-7), abundant in Asia throughout the middle and late Eocene, the cat-like carnivorans (figure 4.11-6) known as nimravids (which evolved into the saber-toothed *Hoplophoneus* and *Nimravus*, extraordinarily convergent on true sabercats of the Ice Age), and even a North American pangolin, or scaly anteater (now found only in the Old World tropics). More important, a number of Chadronian genera apparently evolved *in situ*. These endemics are by far the dominant elements of the White River Chronofauna. They included the rabbit *Palaeolagus*, the oreodont *Merycoidodon* (figures 3.20B, 4.11-3) and the camel *Poebrotherium* (figure 4.11-2).

In summary, the late Eocene forests of North America were invaded by a number of Asian immigrants, which filled the void left by the Uintan extinctions. By the latter part of the late Eocene, the addition of rhinos (figure 4.11-11), peccaries, entelodonts (4.11-1), hyaenodonts (figure 4.11-8), beardogs, cat-like nimravids, and rodents such as squirrels, beavers, pocket mice, pocket gophers, and eutypomyids, gave the landscape a recognizably modern flavor. These animals, along with the abundant horses (figure 4.11-5), camels, brontotheres, rabbits, and especially oreodonts, indicate an ecological association that was adapted mostly for browsing in an open woodland with abundant bushes and restricted grassy and herbaceous openings in the forest (Retallack 1983a, 1983b).

Given the high level of exchange between North America and Asia, it is not surprising that the two continents had many mammals in common in the late Eocene. As discussed in chapter 2, however, confusion over correlation hampers our understanding of the changes in Asian land faunas. Because radiometric dates are scarce, Asian mammals have been correlated to the Eocene and Oligocene by comparison to North American faunas. Now that North American Uintan and Duchesnean are considered middle Eocene, rather than late Eocene, and Chadronian is late Eocene rather than early Oligocene, I have suggested (in Berggren and Prothero 1992) that the Asian sequence also needs to be recorrelated. The only currently available potassium-argon dates (of 31.3 and 32.0 million years on the "middle Oligocene" Hsanda Gol Formation of Mongolia) suggest that the Hsanda Gol and similar Chinese faunas are early Oligocene, as I would predict. New radiometric dates are being measured as I write, so the accuracy of this prediction can be tested soon.

If the Asian "early Oligocene" is actually late Eocene, and if the "middle Oligocene" is actually early Oligocene, then the changes in Asian mammals parallel those seen on other continents. Late Eocene ("early Oligocene" of Li and Ting 1983; Russell and Zhai 1987; and Wang 1992) faunas in Asia included the last of some early Eocene groups that were already extinct elsewhere in the world. These included the bear-like hoofed mesonychids, and the last of the pantodonts— a cow-size beast with huge tusks known as *Hypercoryphodon*. Asia was also home to many mammals characteristic of the Duchesnean and Chadronian in North America. Pig-like entelodonts and hippo-like anthracotheres were the common artiodactyls, along with a variety of deer-like ruminants. Perissodactyls were at their peak in the late Eocene of Asia. Along with a variety of archaic tapirs were huge hippo-like amynodont rhinos (including the appropriately named *Gigantamynodon*) and a variety of long-legged running hyracodont rhinos (some of which reached elephantine size). One genus, *Forstercooperia*, is known from both the Duchesnean of New Mexico and the Shara Murun of Asia. There were also primitive members of the true rhinoceros family and bizarre clawed perissodactyls known as chalicotheres.

As in North America, however, the most spectacular beasts in the Asian late Eocene were the brontotheres. Like North American brontotheres, Asian species have not been thoroughly studied since the classic papers of Walter Granger and William King Gregory in 1943. The Chinese have many new specimens that will require careful reassessment to determine how many species and genera are truly valid. Nevertheless, there were a variety of paired-horned brontotheres that closely resemble North American Chadronian taxa. The most bizarre of all were the embolotheres, which had a single blunt meter-long "battering ram" on the tip of the snout (figure 4.12).

In contrast to North America and Asia, the late Eocene mammal faunas of Europe were much like the rest of the

FIGURE 4.12. The Mongolian "battering ram" brontothere *Embolotherium*. (From Osborn 1929; courtesy U.S. Geological Survey.)

Eocene. Minor changes notwithstanding, the dominant groups continued to be endemic perissodactyls (such as palaeotheres and lophiodonts), artiodactyls, and rodents. Jeremy Hooker (Collinson and Hooker 1987, fig. 10.6) depicted a return of the fruit-eating, squirrel-like pseudosciurid rodents during the late Eocene at the expense of leaf-eating theridomyids and glirids, which had first become abundant at the middle-late Eocene transition. However, the pseudosciurid rebound was short-lived, and glirids and theridomyids again came to dominate the latest Eocene in Europe. There are no such trends in the larger browsing hoofed mammals, or in the overall mammal faunas.

According to Margaret Collinson (Collinson and Hooker 1987), European floras do not show fluctuations during the late Eocene as extreme as those of North America. As noted earlier, Europe was much more humid, and most of the floras are indicative of moist coastal regions in the Eocene archipelago. Nevertheless, the beginning of the late Eocene contains the last record of many tropical plants typical of the earlier Eocene. But the overall climatic decline seems to be a stronger signal; the data are insufficient to ascertain a distinct cooling at the middle-late Eocene transition, followed by a late Eocene warming.

In summary, the late Eocene was a world in transition. Overall biotic diversity was still low after the extinctions at the end of the middle Eocene, but temperatures and floras recovered somewhat, as climatic conditions were not that different from the middle Eocene. The Eocene/Oligocene boundary, as currently defined, was an unremarkable moment in this generally quiet three million years of earth history. There was no spectacular "Terminal Eocene Event," despite a decade of misunderstanding fostered by the term. It was not until the earliest Oligocene (as currently defined) that all hell breaks loose, and we see the true beginning of modern climates and our "icehouse" world.

FIGURE 5.1. Glaciers and icebergs near Chilean base Bernardo O'Higgins on the northwest side of the Antarctic Peninsula. (Photo courtesy R. H. Dott, Jr.)

5

The Big Chill
The Oligocene

And now there came both mist and snow
And it grew wondrous cold
And ice, mast-high, came floating by
As green as emerald

—SAMUEL TAYLOR COLERIDGE, *RIME OF THE ANCIENT MARINER* (1798)

When ratios of oxygen isotopes in the shells of marine organisms were first used as thermometers for the Pleistocene, geochemists assumed that there were no extensive Cenozoic ice sheets before the middle Miocene, about 15 million years ago. Paleotemperatures for the Paleogene were thus calculated using this "ice-free" assumption (Shackleton and Kennett 1975), producing strange results. If continental ice did not cause the high levels of ^{18}O in sea water, then cooler ocean temperatures alone were responsible. And this would indicate equatorial surface waters as cold as 18°C (64°F). But marine animals which lived in those tropical seas, such as corals and benthic foraminiferans (Adams et al. 1990), had temperature tolerances that required conditions no lower than about 28°C (82°F). In addition, the "ice-free" assumption produces estimates of Oligocene bottom water temperatures even lower than the near-freezing values of today. Robley Matthews and Richard Poore (1980) therefore suggested that the ice-free assumption might not be valid for all of the Paleogene. A number of scientists (Miller and Fairbanks 1983, 1985; Keigwin and Keller 1984; Shackleton 1986; Keigwin and Corliss 1986; Prentice and Matthews 1988) have since calculated paleotemperatures assuming at least some continental ice on Antarctica beginning in the early Oligocene. This brought the paleotemperature estimates more in line with other data.

In almost all the oxygen isotope records, the most dramatic change occurred in the earliest Oligocene (Kennett and Shackleton 1975; Miller et al. 1987; Miller 1992; Zachos et al. 1992). Both benthic and planktonic foraminiferans show a brief positive excursion of about 1.3‰ (figure 3.2). Assuming some Oligocene ice, Miller (1992) calculated that 0.3–0.4‰ of the change is due to an ice volume increase, which lowered global sea level by 30 meters (Haq et al. 1987). The remaining 0.9–1.0‰ could then be explained by about 5–6°C (9–11°F) of cooling, lowering global mean temperature to about 5°C (41°F). This contrasts with global mean temperatures as high as 13°C (55°F) in the early Eocene, and 7°C (45°C) in the latest Eocene (Miller et al. 1987). This would have been by far the most dramatic temperature shift in the entire Cenozoic until a similar increase in the middle Miocene signaled the onset of the modern East Antarctic ice sheet.

As discussed in chapter 3, evidence of middle Eocene mountain glaciers on the Antarctic Peninsula (Birkenmajer 1987) and in the Pacific sector of the Southern Ocean (Margolis and Kennett 1971; Wei 1989) suggests a short-term glaciation during the middle Eocene. In 1987 Leg 113 of the ODP, or Ocean Drilling Project (successor to the Deep Sea Drilling Project) drilled the East Antarctic margin, Maud Rise, and the Weddell Sea in the South Atlantic sector of Antarctica and the Southern Ocean (Kennett and Barker 1990). These drill cores produced evidence of ice-rafted detritus (sediments dropped from melting icebergs) beginning in the middle Oligocene (figure 5.2). Dramatic confirmation of early Oligocene ice in Antarctica came in 1988, when Leg 119 of the ODP drilled cores on the submerged Kerguelen Plateau and in Prydz Bay,

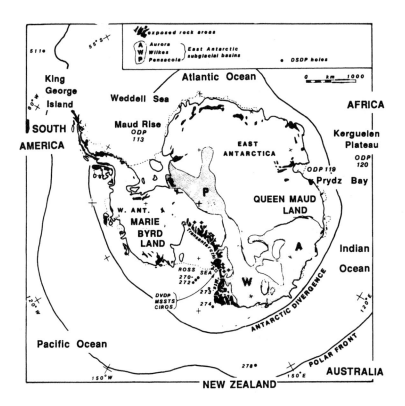

FIGURE 5.2. Map of Antarctica, showing major localities mentioned in text.

both on the Indian Ocean side of Antarctica (Barron et al. 1988). Thick glacial deposits in Prydz Bay and ice-rafted sediments on the southern Kerguelen Plateau showed that a major ice sheet was grounded over part of Antarctica by at least the earliest Oligocene. In addition, glacial sediments were found in even older strata, which suggested a middle or late Eocene glaciation near Prydz Bay, although the age of these deposits is not well constrained. The next leg of the Ocean Drilling Project (Leg 120) drilled the central Kerguelen Plateau and provided further evidence of early Oligocene ice sheets (Breza et al. 1989; Schlich et al. 1989; Zachos et al. 1992). The CIROS-1 drill hole in the Ross Sea also encountered lowermost Oligocene glacial sediments (Barrett et al. 1989). Although the size and duration of these early Oligocene ice sheets is still controversial (Kennett and Barker 1990), it is clear that there were significant ice sheets on some parts of the Antarctic continent about 33 million years ago, and that there were short-term glaciation events since the middle Eocene. Zachos et al. (1992) argued that the evidence suggests a full-scale continental ice sheet on Antarctica for several hundred thousand years.

As we saw in the previous chapter, this oxygen isotope shift had long been attributed to the "Terminal Eocene Event." However, better oceanic cores with a more complete record of the early Oligocene show that most of the increase takes place within magnetic Chron C13N (Miller et al. 1991; Miller 1992)—which is actually earliest Oligocene (about 33 million years old), if we take the *Hantkenina* datum (located in the upper third of Chron C13R, about 34 million years) as the marker of the end of the Eocene (Berggren et al. 1992). Most of the events traditionally ascribed to the "Terminal Eocene Event" take place about a million years after the Eocene/Oligocene boundary. The term *Terminal Eocene Event* coined by Wolfe (1978) came from miscorrelations of the early Oligocene floral change with the Eocene/Oligocene boundary; the actual faunal and climatic changes at the "Terminal Eocene Event" were pretty minor. Perhaps a better label would be Wolfe's (1971) *Oligocene deterioration*.

The dramatic cooling and ice buildup beginning in the Oligocene were accompanied by many other changes. Deep-sea cores show unconformities in the earliest Oligocene, indicating vigorously circulating deep waters in the Southern Ocean (Kennett 1977) and even in the Arctic and northern North Atlantic (Miller and Tucholke 1983). Apparently, there were bipolar sources for cold deep waters. By comparing the carbon isotope ratios in different oceans in the earliest Oligocene, Miller (1992) saw evidence of a pulse of cold, nutrient-depleted waters from the North Atlantic, and a cold but moderately nutrient-rich pulse from the Antarctic. These pulses increased circulation and ventilation of the world's oceans, but they eroded deep-sea records. A striking feature of the carbon isotope ratios is that they are similar in all the oceans; they do not show the basin-to-basin differences that are typical of other times. Traditionally, this homogeneity has been explained as reflecting a single source (the Antarctic) for all deep waters, but Miller (1992) suggested that it might also reflect nutrient depletion in the lower latitudes during the early Oligocene. With reduced organic carbon in a depleted ocean, the isotopic shifts became less visible and tend to be masked by "noise." However, the nutrient depletion explanation cannot be applied to the Southern Ocean, which experienced upwelling and nutrient-triggered planktonic blooms in the early Oligocene (Aubry 1992; Baldauf 1992).

Boersma et al. (1987) described planktonic foraminiferans of the early Oligocene as a homogeneous fauna that was small in size, low in diversity, and found over much of the world (figure 5.3). This indicated that the cold oceans were well circulated, not stratified by depth or temperature. The abundance of deep-water, cold-tolerant biserial heterohelicid foraminiferans (figure 5.3A) is particularly diagnostic of the shoaling and mixing of cold bottom waters with the surface. Only the tropical zone retained some stratification, as indicated by the difference in carbon isotopic values between surface- and deep-dwelling foraminiferans.

Even so, the Mediterranean was apparently isolated from the Atlantic; its early Oligocene foraminiferans and isotopic values became increasingly different from that of the Atlantic. This

171

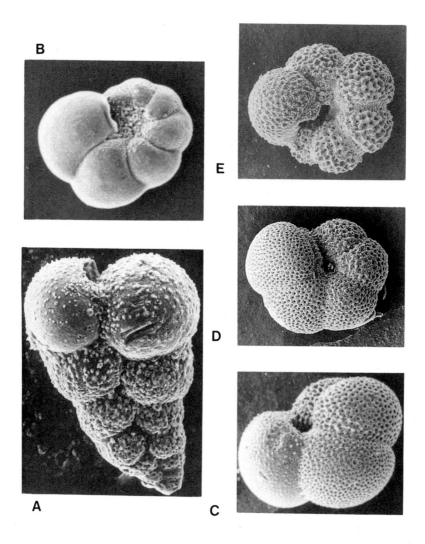

FIGURE 5.3. Characteristic Oligocene planktonic foraminiferans. (A) *Chiloguembelina cubensis*, a biserial foram abundant in the early Oligocene. (B) *Pseudohastigerina micra*, index fossil of the earliest Oligocene. (C) *Globigerina ampliapertura*, index fossil of the late early Oligocene. (D) *Globorotalia opima*, index fossil of the mid-Oligocene; it is associated with the cooling of the oceans. (E) *Globigerina ciperoensis*, index fossil of the late Oligocene. Each specimen is about 200-300 microns in diameter. (Photos courtesy G. Keller.)

could have been caused by late Eocene shoaling of the straits between the eastern Mediterranean and the remnant of Tethys across what is now the Middle East (Ricou et al. 1986). As a result, the balance of flow through the Gibraltar passage was altered; less Tethyan water flowed westward into the Mediterranean and out into the Atlantic. Boersma et al. (1987) suggested that the shutoff of Mediterranean deep water from the Atlantic would have decreased stratification in the central Atlantic because cooler waters from the Antarctic and Arctic/North Atlantic would have swept in.

The cooling also affected the temperature-sensitive marine algae. Aubry (1992) contends that the earliest Oligocene cooling triggered high rates of extinction in the coccoliths, especially among temperate forms. At this time, all but one of the long-ranging species that had evolved in the early Eocene, several of the long-ranging taxa that evolved in the middle Eocene in response to climatic deterioration, and all the short-ranging taxa that evolved in the late Eocene became extinct, bringing to 70% the reduction in diversity since the extinctions began in the middle Eocene. Although the overall extinction rate was high, cold-tolerant coccoliths bloomed in huge numbers due to the release of nutrients by vigorous deep-water currents. By the late early Oligocene, the segregation of high-latitude cold water masses in the Antarctic led to increased provincialism, as cold-tolerant species also became restricted to the Southern Ocean cold water masses. According to Baldauf (1992), about 45% of the Eocene diatom species became extinct by the end of the early Oligocene, and they were replaced by almost the same number of new species. Although all latitudes experienced this turnover, the most striking change was the great increase in siliceous productivity in the Antarctic. Extinction and speciation were probably related to the increase in deep-water currents, which cycled deep-water nutrients back into the surface waters (Baldauf and Barron 1990). There were also significant extinctions of dinoflagellates at the beginning of the Oligocene (Brinkhuis 1992).

Benthic foraminiferans, too, show dramatic changes. In the Antarctic, Thomas (1992) concluded that earliest Oligocene assemblages had dramatically lower faunal diversity than those of the Eocene, with the deep-water *Nuttalides umbonifera* composing up to 70% of some assemblages. Foraminiferans that lived on the ocean bottom decreased, and those living within the sediment increased, suggesting that the ocean chemistry had become corrosive to calcite-shelled organisms as deep, cold waters began their rapid circulation. In the U.S. Gulf Coast, there were also high rates of extinctions among benthic foraminiferans (Gaskell 1991). Similar extinctions took place among the last survivors of the warm-water Tethyan province, including many of the disc-shaped nummulitids (Adams et al. 1986). Agglutinated foraminiferans, which build their shells from cemented sand grains, were also affected (Kaminski 1988). They became extinct sequentially from south to north, perhaps because cold corrosive Antarctic bottom waters eliminated agglutinated taxa as the currents flowed northward. By the early Oligocene, agglutinated foraminiferans had disappeared from the North Atlantic and other areas, surviving only in the isolated Norwegian-Greenland Sea.

Larger benthic organisms were also victimized. After the severe extinctions of molluscs at the end of the middle Eocene (Hansen 1987), another pulse of molluscan extinction occurred in the early Oligocene. In the U.S. Gulf Coast, about 97% of the gastropods and 89% of the bivalves that survived the middle Eocene extinction did not make it into the late early Oligocene (Hansen 1987, 1992). As with middle Eocene extinctions, warm-water molluscs were the chief victims. Although echinoids did not suffer a dramatic extinction at the end of the middle Eocene, they declined dramatically during the early Oligocene cooling, when over 50% of the echinoid species became extinct worldwide (McKinney et al. 1992). The importance of cooling to this extinction is demonstrated not only by the disproportionate extinction of tropical echinoids but also by the evolution of cold-adapted marsupiate echinoids in the late Eocene of Australia. They managed to survive and spread during the colder Oligocene.

At the top of the marine food pyramid are the great whales. Eocene archaeocete whales were primitive beasts, similar in many ways to their mesonychid ancestors. Most archaeocetes are known from the lower latitudes, particularly the U.S. Gulf

Coast and the Tethyan belt, indicating their preference for warmer waters. But archaeocetes disappear completely at the end of the Eocene. In their place evolved the earliest representatives of the two modern groups of whales: the toothed whales (such as the sperm whale, killer whale, and dolphins), and the baleen whales (such as the blue whale, right whale, gray whale). Baleen whales strain microscopic food particles (plankton, crustaceans, and small fish) out of the water with a filter of baleen in their mouths. The first baleen whales are known from the earliest Oligocene of the Southern Ocean region (primarily Australia and New Zealand), and the first toothed whales from the late early Oligocene of the same region (Fordyce 1992). By the late Oligocene, there were as many as fifty species in more than eleven families. Fordyce (1980, 1989, 1992) argues that this explosive diversification is directly related to the increase in marine plankton around Antarctica as renewed deep-water circulation of the early Oligocene released nutrients and triggered an explosive plankton bloom. Indeed, some of the densest population of whales today are found in the plankton-rich waters around Antarctica.

The Earliest Oligocene Climatic Deterioration in North America

One of the most dramatic responses to the early Oligocene cooling was demonstrated by North American land plants. In the middle Eocene, continental climate exaggerated the global temperature change, producing much more extreme chilling on land. The Oligocene deterioration, however, brought the most extreme paleobotanical change in all of the Cenozoic. Based on his method of leaf margin analysis, Wolfe (1971, 1978, 1985, 1992) suggested that mean annual temperatures cooled about 8–12°C (13–23°F) in less than a million years (figures 3.7, 5.4). In Alaska, for example, the pre-deterioration Rex Creek floral assemblage (figures 5.5A, B) was a broadleaved evergreen forest, with a preference for mean annual temperature conditions of about 15°C (59°F), and a range of no more than 10°C (18°F). Typical plants included the conifer *Engelhardtia*, holly, and Labrador tea. By contrast, Alaskan floras after the deterioration (figure 5.5C) indicate a mixed northern hardwood forest composed of alders, beeches, hickories, and dawn redwoods

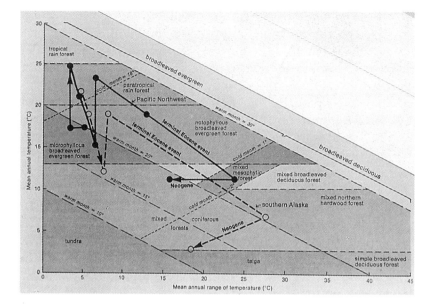

FIGURE 5.4. Plot of mean annual temperature versus mean annual range of temperature for different climatic and vegetational types. The fluctuations at the middle-late Eocene boundary, and the Oligocene deterioration (here labeled "terminal Eocene event") show up in the trajectory of forest types from the Pacific Northwest and southern Alaska (Wolfe 1978; by permission of *American Scientist*).)

(*Metasequoia*), tolerant of a mean annual temperature of 7°C (44°F). This suggests an 8°C (15°F) drop in mean annual temperature! Even more striking, these floras could tolerate temperatures as much as 28°C (50°F) above and below the mean, indicating severe freezing conditions in the winter and hot summers as now occurs in northern hardwood forests of eastern Canada and New England.

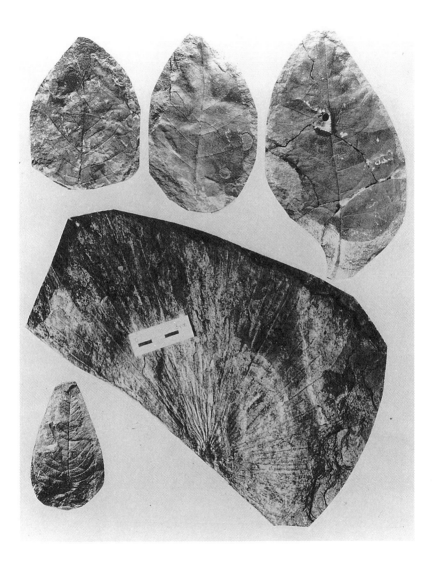

FIGURE 5.5A. Comparison of pre-deterioration (late Eocene) and post-deterioration (early Oligocene) leaf fossils from Alaska. Late Eocene floras include the tree ferns *Anemia* (center) and *Allantodiopsis*. (From Wolfe 1977; courtesy U.S. Geological Survey.)

FIGURE 5.5B. Other characteristic tropical elements of the late Eocene include the palm *Palmacites* (center) as well as alders (top three leaves) and the pecan tree *Platycarya*. (From Wolfe 1977; courtesy U.S. Geological Survey .)

FIGURE 5.5C. Post-deterioration leaves are all much smaller, and many have jagged margins. These examples include poplar, alder, willow, walnut, winter hazel, holly, and plum trees. (From Wolfe 1977; courtesy U.S. Geological Survey.)

THE BIG CHILL

The change was no less dramatic in the Pacific Northwest. In western Oregon (figure 3.10), the latest Eocene Comstock flora (Sanborn 1935) and earliest Oligocene Goshen flora (Chaney and Sanborn 1935) were both paratropical rain forests (Wolfe 1978), indicating mean annual temperatures of about 20–22°C (68–72°F), about 180 cm (70 inches) of annual rainfall, and tolerating a range of temperatures of only 7°C (13°F) above and below the mean. Goshen plants include oaks, figs, magnolias, paw-paws, holly, soapberry, laurels, myrtles, ebonies, roses, legumes, and herbs such as borages and heliotropes, as well as a number of less familiar subtropical families typical of the temperate rain forest of Costa Rica. Other latest Eocene-earliest Oligocene predeterioration floras include the Upper Clarno in central Oregon (figure 3.10B), and Sweet Home (Brown 1950) and Bilyeu Creek/Scio floras (Sanborn 1949) in western Oregon. Each of these assemblages is overlain by an early Oligocene, postdeterioration flora characteristic of mixed broadleaved decidous forests. For example, the Goshen flora is overlain by the Willamette, the Bilyeau Creek/Scio by the Lyons, and the Upper Clarno by the Bridge Creek. Typical postdeterioration plants include (in the case of Bridge Creek) oaks, laurels, alders, dawn redwoods, ash, hawthorns, mock orange, sycamores, walnuts, hackberries, elms, lindens, beeches, bracken ferns and horsetails. In many ways, these floral assemblages are similar to modern redwood forests. Available dates show that the timespan of deterioration was in the order of a million years or less. According to Wolfe (1978), the mean annual temperature in post-deterioration forests of Oregon was about 12°C (54°F), a decline of at least 10°C (18°F) since the earliest Oligocene, and there was a much greater range of seasonal temperatures (approaching 24°C, or 43°F above and below the mean).

Although many of the floras in the Rocky Mountains were more strongly affected by altitude than by the Oligocene climatic deterioration (Wing 1987), some of these floras clearly show the cooling trend (Wolfe 1992). For example, the famous Florissant lake beds of the late Eocene in central Colorado (MacGinitie 1953) preserved a high-altitude (2450 meters, or 8000 feet—Meyer 1986) streamside forest of firs, spruce, pines,

FIGURE 5.6. Fossilized giant sequoia stumps from the earliest Oligocene (pre-deterioration) Florissant Fossil Beds National Monument, high in the Rocky Mountains west of Colorado Springs, Colorado. Note the people on the bridge for scale. A few miles to the west of this locality, the slightly younger (post-deterioration) Antero flora contains a cold-adapted northern hardwood forest, and no giant sequoias.

cypress, sequoias (figure 5.6), elms, hackberries, cottonwoods, mountain mahogany, hawthorn, apples, plums, with mosses, cattails, and horsetails as well. According to Wolfe (1992), this forest was characterized by a mean annual temperature of 12.5°C (55°F), and an annual range of temperatures of about 24°C (43°F)—much like the post-deterioration floras of the Pacific Northwest. However, the early Oligocene Antero flora of Colorado is only about half a million years younger than the Florissant, and it is very similar to the late Oligocene Creede flora of Colorado. All these assemblages include plants typical of a northern hardwood forest (pines, oaks, blueberries, mountain mahogany, and some subalpine plants), indicating a mean annual temperature of 4.5°C (40°F). Thus, even high-altitude regions in the Rocky Mountains experienced about 8°C (15°F) of cooling.

181

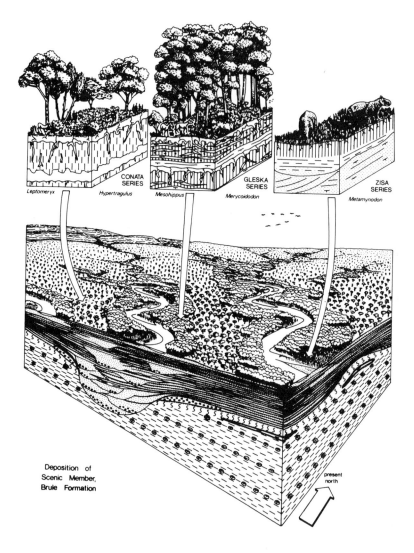

FIGURE 5.7A. Reconstruction of late Eocene (Chadronian) landscapes in the Big Badlands, showing the dense woodlands indicated by the paleosols. (From Retallack 1983; by permission of the Geological Society of America.)

182

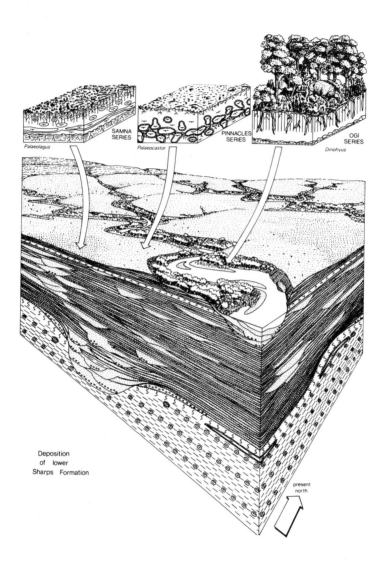

FIGURE 5.7B. Late Oligocene (early Arikareean) landscapes, showing more open shrub and grasslands with streamside woodlands. (From Retallack 1983; by permission of the Geological Society of America.)

183

THE BIG CHILL

In the U.S. Gulf Coast, late Eocene (Jackson Formation) floras represent tropical rain forests with a mean annual temperature of about 28°C (82°F). Unfortunately, the overlying lower Oligocene Vicksburg Formation produces too few megascopic plants for comparison. There are, however, striking changes in the pollen (Frederiksen 1980). In Mississippi and Alabama are found massive extinctions of late Eocene pollen taxa; Vicksburg pollen include a great abundance of cool, dry-adapted oaks. In a core from Mays Landing, New Jersey, Ager (in Owens et al. 1988) found typical Jackson-type pollen into the earliest Oligocene. Only 3 meters higher in the core, another sample showed a dramatic shift to predominantly oak-like pollen in the early Oligocene. The Oligocene climatic deterioration was clearly a continent-wide feature.

Even in regions where organic materials (including pollen or megascopic plant fossils) are not well preserved, there is evidence of vegetational change. In the Big Badlands of South Dakota (which produces many of the critical mammal fossils), the only plant fossils are hackberry seeds (which are virtually fossilized with calcite even when alive), and fossil wood of walnut trees, so the fossil flora is too small for paleoclimatic reconstruction. However, Greg Retallack (1983a, 1983b, 1990, 1992) concluded that ancient soil horizons (paleosols) preserved in the upper Eocene Chadron Formation were formed under moist closed forest canopies, with between 500–900 mm (20-35 inches) of rainfall per year (figure 5.7A). In the overlying strata, the lower Oligocene (Orellan) Scenic Member of the Brule Formation contains paleosols indicating less than 500 mm (20 inches) of rainfall per year, and the soil structure is typical of a more open, dry woodland. In eastern Wyoming, Emmett Evanoff (in Evanoff et al. 1992) found that moist Chadronian floodplain deposits abruptly shifted to drier, wind-blown deposits in the early Orellan. These same beds preserve an excellent record of climate-sensitive land snails. According to Evanoff, Chadronian land snails are large-shelled taxa similar to snails now found in subtropical climates with seasonal precipitation, such as in the southern Rocky Mountains and central Mexican Plateau. Based on modern analogues, these fossils indicate a mean annual temperature of 16.5°C (62°F)

FIGURE 5.8. Land tortoises are among the commonest fossils in the Orellan, in contrast to the aquatic turtles and crocodiles found in older beds. The Scenic Member of the Brule Formation used to be known as the "Turtle-*Oreodon*" beds. (From Leidy 1853.)

and a mean annual precipitation of about 450 mm (18 inches) during the late Chadronian in eastern Wyoming. By the Orellan, the large-shelled snails had been replaced by drought-tolerant, small-shelled taxa indicative of a warm-temperate open woodland with a pronounced dry season. Such snail faunas are today found in southern California and northern Baja California.

The amphibian and reptile fauna shows a similar trend toward cooling and drying. According to Hutchison (1982, 1992), the aquatic forms (especially salamanders, freshwater turtles, and crocodilians) steadily declined in the late Eocene, and by the Oligocene only terrestrial tortoises (figure 5.8) were common. This indicates a pronounced drying trend during the late Eocene. Crocodiles were absent by the Chadronian, but more cool-tolerant alligators persist until the early Orellan.

A

B

FIGURE 5.9. Characteristic Oligocene outcrops of the White River Group in the High Plains. (A) Exposures of the Brule Formation in Badlands National Park, South Dakota. The ledgy sandstones of the Scenic Member at the bottom represent well-watered floodplains, but the massive wind-blown deposits of the Poleslide Member at the top suggest a much more arid climate. (B) The Chadronian-Orellan transition looking south from Douglas, Wyoming. The boundary lies just above the white ash bed (PWL) midway up the cliff (see figure 2.9). The Laramie Range appears in the distance.

186

The abundant and well-preserved mammals of the Big Badlands (and other outcrops of the White River Group from North Dakota to Colorado) are the most famous fossils of the Eocene-Oligocene deterioration in North America (figure 5.9). Despite the dramatic changes seen in the plants, soils, snails, and lower vertebrates, only a small drop in mammalian diversity marked the Chadronian-Orellan transition (figure 5.10) (Prothero 1985; Stucky 1990, 1992). A few more of the archaic elements so characteristic of the Eocene forests became extinct during the Chadronian. The last of the multituberculates died out in the mid-Chadronian, about 35 million years ago. But several other archaic groups of rodents and insectivores, along with archaic relatives of camels, and the mole-like epoicotheres, disappeared during the Chadronian-Orellan transition. The most spectacular victims were the huge brontotheres (figure 5.10). Osborn (1929) argued that brontotheres died out because their internally-driven evolution toward large size and huge horns was out of control. It is much simpler to argue that brontotheres (which had extremely low-crowned teeth suitable only for eating soft leafy vegetation) died out, along with most of the other archaic forest browsers, as a consequence of the drying and breakup of the Chadronian forests.

Accompanying the extinctions of some archaic mammals was the appearance in the Orellan of a few rodents and artiodactyls with high-crowned teeth well-suited for eating tough vegetation. The bulk of the White River Chronofauna, however, changed very little, despite clear evidence of a dramatic climatic and vegetational change. As in the Chadronian, the most abundant Orellan mammals were oreodonts (figure 3.20B), plus the three-toed horses *Mesohippus* and *Miohippus* (figure 4.11-5), primitive camels (figure 4.11-2), the long-legged hyracodont rhinos (figure 4.11-9), and a variety of deer-like artiodactyls. In the streams were the last of the aquatic amynodont rhinos, as well as primitive members of the true rhinoceros family (figure 4.11-11) and the hippo-like anthracotheres (figure 4.11-7). For most of these mammals, the same genera and often the same species found in the late Chadronian are also found in the Orellan. This puzzling fact suggests that animals are not infinitely plastic in their evolution.

187

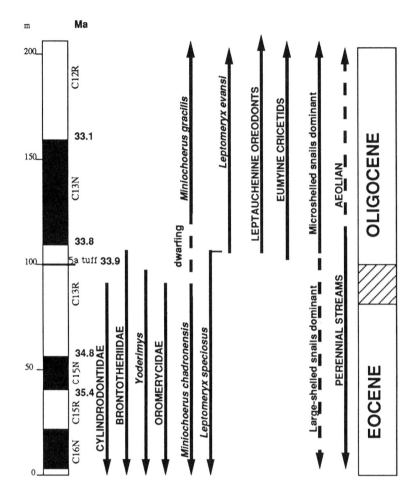

FIGURE 5.10. Patterns of change at the Chadronian-Orellan transition near Douglas, Wyoming (see figure 5.7B). The land snails and sedimentary rocks (right) show a drying trend from perennial streams to sand dunes, suggesting a severe climatic shift. However, the response of the mammals is remarkably mild. A few groups of archaic mammals (left) go extinct, including the rhino-like brontotheres, the camel-like oromerycids, and several archaic rodent groups. The small oreodont *Miniochoerus* dwarfs by about a third through the transition. Two new groups of mammals, the leptauchenine oreodonts (see figure 5.20) and the cricetid rodents, with high-crowned teeth for tough vegetation, appear in the Orellan. Most of the hundreds of species of mammals, however, do not change despite the radical transformation in climate.

188

In the face of dramatic climatic change, we do not see rapid evolutionary changes in the mammals. Most mammalian species were static over millions of years, and only a few responded to the environmental change (either by becoming extinct, or by changing in relative abundance). Only one lineage (a small oreodont known as *Miniochoerus*) evolves significantly in response to the drying. Over about a million years during the Chadronian-Orellan transition, this oreodont gradually dwarfed to about two-thirds its pre-deterioration size.

La Grande Coupure

One of the most important chemicals used in fertilizers and gunpowder is phosphate. When an abundant source of phosphate is discovered, it is quickly exploited. Much of the world's concentration of phosphate comes from animal sources. Bird droppings and other fecal matter are rich in phosphate. In some places, abundant droppings of fish-eating seabirds (known as guano) are so encrusted on sea cliffs and islands that the guano can be mined commercially. Bat guano has been mined from caves. Vertebrate bone is made mostly of phosphate, so abundant sources of bone can serve as fertilizer. In some coastal areas, ground-up fish meal is used.

In some parts of the world phosphate occurs in pockets of ancient limestone. One such deposit is located in the Quercy region of south-central France. First discovered in 1865, it was mined for many years before its true value was realized. Much of the phosphate in the Quercy mines was in the form of actual vertebrate bones, and they were remarkably well preserved. Each pocket of phosphate occurred in what had been a cavern or fissure dissolved out of the ancient limestone bedrock during the Eocene and Oligocene. These fissures accumulated bones from animals that fell or were swept into them 40 to 30 million years ago. The quality of the fossils rivals the great finds of Cope, Leidy, and Marsh in North America. In the introduction to his 1876 monograph on the Quercy mammals, the French paleontologist Henri Filhol enthused,

The localities of Quercy must be considered as having yielded the most interesting evidence hitherto discovered in Europe for the study of fossil mammals, and the animal forms which they reveal are no less valuable than those which have been brought to light in America in recent years.

Unfortunately, fissures opened at different times, so their entombed treasures are all of different geologic ages. At first this jumble of fossils confused the scientists, who assumed that everything from the Quercy pockets was the same age.

After 1890 the commercial value of the French phosphate plummeted, since much cheaper phosphates (also a source of vertebrate bones) were found in North Africa. The Quercy mines were abandoned. In the 1960s scientists from the universities in Montpellier and Paris began to systematically reopen the old mines. Because nobody knew which pocket (and therefore which age) the fossils in the old collections had come from, it was necessary to reexcavate each source and sort out the complex sequence of ages. Over the last few decades a time sequence for more than a hundred localities and their fossils was determined. Many more fossils (especially small mammals) that were undiscovered or neglected by the nineteenth-century scientists have been unearthed.

The Quercy deposits contain a staggering array of opossums, carnivorans, hyaena-like creodonts, bats, rodents, lemur-like primates, and a great variety of endemic perissodactyls (palaeotheres and lophiodonts) and artiodactyls. As noted earlier, most of the middle and late Eocene mammals of Europe were endemic to their islands, isolated by high sea levels (figure 3.25). There was very little exchange between western Europe (from southern England to Germany to Quercy in France) and any other region.

Since the turn of the twentieth century, however, European vertebrate paleontologists have noticed a striking change in the sequence of faunas at Quercy, in northern France and Belgium, and in England. As early as 1909, Swiss paleontologist Hans Stehlin referred to this change as *la Grande Coupure* ("the great break"). Sixty percent of the typical late Eocene mammals (dominated by endemic perissodactyls and artiodactyls, primates, creodonts, archaic insectivores, and browsing

theridomyid rodents) became extinct. In their place, the post–
Grande Coupure mammals (clearly immigrants from elsewhere) included some of the earliest members of the bear,
raccoon, weasel, and mongoose families, as well as the now-
extinct beardogs and cat-like nimravids. Primates were rare,
and archaic insectivores were gone. Although some of the
endemic hoofed mammals persisted, most were displaced by
immigrants. Among perissodactyls, the last of the palaeotheres
were joined by true rhinoceroses, fleet hyracodont rhinos,
aquatic amynodont rhinos, and clawed chalicotheres. Surviving
endemic artiodactyls had to compete with immigrant pig-like
entelodonts, the earliest peccaries, hippo-like anthracotheres,
and a variety of deer-like ruminants. Theridomyid rodents also
persisted, but they were largely displaced by immigrant
families, including the aplodontids (represented by the sewellel,
or "mountain beaver"), true beavers, squirrels, glirids (dormice)
and cricetids (hamsters, New World mice, and their relatives),
plus the rabbits.

As on other continents, a striking difference between older
and younger faunas is the disappearance of arboreal mammals
(especially primates), and the replacement of leaf-eaters with
grain-eaters and other herbivores adapted to tougher vegetation. Another big difference can be seen in the distribution of
mammalian body sizes. As noted in chapter 1, the cenogram of
body sizes is indicative of habitat (figure 1.9). In the case of the
post–Grande Coupure mammals, Legendre (1987, 1988) plotted
a cenogram with a much steeper slope rising to the large-size
portion of the plot (thus indicating a greater biomass of large
hoofed mammals). Legendre's cenogram also has a sharp
break in the middle size range, and then a smaller range of
sizes for medium to small mammals (figure 5.11). This distribution is characteristic of a cooler, drier, more open and
savanna-like habitat, than that coincident with the Eocene
rainforests.

Where did all the immigrants come from? There had been no
land route across the Atlantic since the early Eocene, so they
had to come from Asia, or from North America via Asia

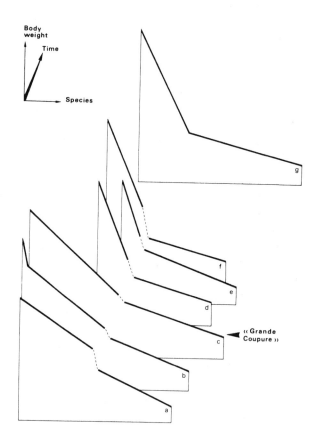

FIGURE 5.11. Cenogram of European mammalian faunas from the Phosphorites de Quercy, France, showing the sharp change in size distribution after the "Grande Coupure." Before that event, the cenograms suggest a tropical rainforest, but afterward they are comparable to curves from modern savannas or deserts. Compare with figure 1.9A. (Modified from Legendre 1988.)

(Matthew 1906; Stehlin 1910; McKenna 1983; Savage and Russell 1983). The "Grande Coupure" is thus less a climatic event and more an example of sudden immigrant influx with displacement of native endemic taxa (comparable to the Pliocene "Great American Interchange" between South and North America). This immigration, however, may have had climatic triggers. Conventionally, the flood of Asian and North American taxa was thought to have crossed the Turgai Straits across the Ural region, which separated Asia and Europe in the Eocene (Vianey-Liaud 1976). A lowering of sea levels in the early Oligocene may have dried up these seaways and opened land passages (Plint 1988). However, Heissig (1979) pointed out that some of these immigrants (such as brontotheres and anthracotheres) had already migrated to the Balkan region in the middle Eocene. He argues for a more complex corridor system, with the Alpine uplift serving as the main corridor between eastern and western Europe during the Grande Coupure.

The relatively minor role of climate in causing the Grande Coupure is evident in the vegetational changes across the Eocene-Oligocene transition in Europe. Unlike North America, the changes in European floras were rather subtle (Collinson and Hooker 1987; Collinson 1992). The late Eocene evergreens, bald cypresses, and reed marshes were replaced by a mixed deciduous/evergreen flora indicating a warm temperate climate. Reed marshes were again dominant in the early Oligocene, although trees were mostly members of the walnut and elm families. Pollen samples show an increase in temperate plants and conifers and the loss of the last tropical and subtropical vegetation (Hubbard and Boulter 1982; Boulter and Hubbard 1983; Boulter 1984; Collinson 1992). There is clear evidence of a cooling trend, but not of the drying trend seen in North America. As noted earlier, floral changes in Europe may have been moderated by the fact that they were located on humid coastal islands, which had a much more moderate climate than the extremes of the North American continent.

Conventionally, the Grande Coupure was correlated with the Eocene-Oligocene boundary (Cavelier 1979; Savage and Russell 1983), and some paleontologists have argued that the

Eocene-Oligocene boundary should be *defined* on the basis of the Grande Coupure. (But the Eocene and Oligocene were originally based on marine invertebrates and sediments, so their boundary cannot be based on mammal faunas). However, a few dissenters (Lopez and Thaler 1974; Sigé and Vianey-Liaud 1979) have suggested that the Grande Coupure might be early Oligocene. Because we now know that the Eocene-Oligocene boundary was relatively uneventful, and that the big change occurred in the earliest Oligocene, this issue is critical. According to Jeremy Hooker (1992), there are several sequences in England and Germany which demonstrate that the Grande Coupure is earliest Oligocene, and clearly not equivalent to the Eocene-Oligocene boundary. If so, then the Grande Coupure probably corresponds to the great sea level drop and ice volume increase in the Antarctic that triggered the North American floral deterioration between 34 and 33 million years ago. This makes sense, because the sea level retreat would have provided corridors for the immigration into Europe of the Asian mammals that largely replaced the endemics.

Asian Giants
As discussed in Chapters 2 and 4, recorrelation of the North American Chadronian with the late Eocene suggests that the Asian "early Oligocene" is also late Eocene. Similarly, Asian faunas of the early Oligocene ("middle Oligocene" of Li and Ting 1983, Russell and Zhai 1987, and Wang 1992) included abundant cricetid rodents (hamsters, New World mice, and their kin), true rhinoceroses, and a variety of mammalian genera (listed in Berggren and Prothero 1992:21) found in the early Oligocene elsewhere. Conspicuous by their absence were typical late Eocene groups such as brontotheres, primitive chalicotheres and tapirs, pantodonts and mesonychids. Many species of amynodont and hyracodont rhinos, and anthracotheres also disappeared. This pattern of extinctions closely resembles the Chadronian-Orellan (Eocene-Oligocene) transition in North America. Indeed, if the entire Asian sequence were correlated as Wang (1992), Li and Ting (1983), Russell and Zhai (1987), and Dashzeveg and Devyatkin (1986) suggest, it would be hard to explain why everything in Asia is out of synch

with the rest of the world and all major extinction events seem to happen three million years later in Asia.

The Hsanda Gol fauna of Mongolia and similar early Oligocene faunas in China include some amazing beasts. Rabbits were abundant, as were a variety of modern rodent families (beavers, sewellels, cricetids, ctenodacytids or gundis, jumping mice, cane rats, and bamboo rats). Predators included the last hyaenodont creodonts (figure 3.21A) and several modern carnivoran families (bears, raccoons, weasels, mongooses) plus the extinct beardogs and cat-like nimravids. The usual entelodonts, anthracotheres, and deer-like ruminants were also present, along with lingering archaic tapirs and amynodont rhinos. The most spectacular beasts were the gigantic hyracodont rhinos known as indricotheres. This group included the largest land mammal the world has ever known, *Paraceratherium* (sometimes called by the obsolete names *Baluchitherium* or *Indricotherium*). This beast was almost 18 feet (6 m) high at the shoulder and probably weighed 20 tons (18,000 kg). Its head was so high off the ground that it browsed on the tops of trees over 25 feet (7.5 m) high (figure 5.12A). Today we think of elephants and giraffes as giants, but *Paraceratherium* dwarfed them in both size and bulk. Its head was over 5 feet (1.5 m) long, with a pair of enormous tusks at the front end of its skull. As big as its head was, it seemed ridiculously small on such a large body.

In spite of these bizarre features, *Paraceratherium* still bears the hallmarks of its hyracodont ancestry. Its molar teeth show the same pattern as the hyracodonts, only they are enormous. The incisors, although large, are conical as they are are in hyracodonts. Significantly, its toe bones are still long and stretched out as if it were a runner. This feature is truly remarkable because other gigantic land animals, such as elephants and dinosaurs, demonstrate a shortening of foot bones until they are stubby, square blocks or even flattened like pancakes. The indricotheres outweighed any elephant, yet they retained the long toes of their running ancestors. An animal this large clearly had no need to run from any predator; it was much too large to run efficiently anyway. *Paraceratherium* is a good

FIGURE 5.12A. Reconstruction of the giant hyracodont rhinoceros *Paraceratherium* (formerly known as *Indricotherium* or *Baluchitherium*) from the early Oligocene in Mongolia. To the right of it is a reconstruction of the running rhino *Hyracodon* from which it was descended, and a modern elephant for scale. (Courtesy University of Nebraska State Museum).

FIGURE 5.12B. Four complete limbs of *Paraceratherium* as they were entombed when the animal sank into quicksand.

example of how animals sometimes retain ancestral features long after such have outlasted their usefulness.

The American Museum Mongolian expeditions of 1922 and later years made a number of spectacular finds, including the first dinosaur eggs. But the gigantic bones of *Paraceratherium* were among the most exciting. Roy Chapman Andrews, the leader of the expedition, described it this way:

> The credit for the most interesting discovery at Loh belongs to one of our Chinese collectors, Liu Hsi-ku. His sharp eyes caught the glint of a white bone in the red sediment on a steep hillside. He dug a little and then reported to [Walter] Granger [the chief paleontologist of the expedition] who completed the excavation. He was

197

amazed to find the foot and lower leg of a *Baluchitherium*, STANDING UPRIGHT, just as if the animal had carelessly left it behind when he took another stride [figure 5.12B]. Fossils are so seldom found in this position that Granger sat down to think out the why and wherefore. There was only one possible solution. Quicksand! It was the right hind limb that Liu had found; therefore, the right front leg must be farther down the slope. He took the direction of the foot, measured off about nine feet and began to dig. Sure enough, there it was, a huge bone, like the trunk of a fossil tree, also standing erect. It was not difficult to find the two limbs of the other side, for what had happened was obvious. When all four legs were excavated, each one in its separate pit, the effect was extraordinary. I went up with Granger and sat down upon a hilltop to drift in fancy back to those far days when the tragedy had been enacted. To one who could read the language, the story was plainly told by the great stumps. Probably the beast had come to drink from a pool of water covering the treacherous quicksand. Suddenly it began to sink. The position of the leg bones showed that it had settled slightly back upon its haunches, struggling desperately to free itself from the gripping sands. It must have sunk rapidly, struggling to the end, dying only when the choking sediment filled its nose and throat. If it had been partly buried and died of starvation, the body would have fallen on its side. If we could have found the entire skeleton standing erect, there in its tomb, it would have been a specimen for all the world to marvel at.

I said to Granger: "Walter, what do you mean by finding only the legs? Why don't you produce the rest?" "Don't blame me," he answered, "it is all your fault. If you had brought us here thirty-five thousand years earlier, before that hill weathered away, I would have had the whole skeleton for you!" True enough, we had missed our opportunity by just about that margin. As the entombing sediment was eroded away, the bones were worn off bit by bit and now lay scattered on the valley floor in a thousand useless fragments. There must have been great numbers of baluchitheres in Mongolia during Oligocene times, for we were finding bones and fragments wherever there were fossiliferous strata of that age (Andrews 1932:279-80)

Paraceratherium was probably as large as a land mammal can become. Only the whales are larger, as their weight is carried by the buoyancy of water. Some people have suggested that indricotheres were also aquatic to help bear their enormous weight, although their bones were certainly robust enough to carry them. In addition, their enormous height and long necks only make sense if they browsed on treetops, as giraffes do. Gigantic herbivores, such as elephants and dinosaurs, must consume enormous amounts of vegetation to support such large bodies. Living elephants must eat almost constantly to survive. Jim Mellett (1982) has shown that *Paraceratherium* was probably a hindgut fermenter (digesting with the intestines), like other rhinos and elephants, and therefore was not as efficient at digestion as ruminants, such as cows or giraffes, which have four-chambered stomachs to serve as microbial vats. Instead, *Paraceratherium* had to pass large amounts of relatively low-quality forage through its gut quickly in order to get sufficient energy from its food intake. The largest dinosaurs, which were four times as big as *Paraceratherium*, had peg-like teeth that could not slice up vegetation. They had to swallow their food whole and digest large amounts of it quickly to survive. *Paraceratherium* was one of the few mammals that made a living like the browsing dinosaurs. Not surprisingly, very few mammals have tried it before or since because it is a very difficult lifestyle in terms of bioenergetics. *Paraceratherium* was the largest land mammal ever seen, and it's unlikely that any mammal will ever top its record.

Late Paleogene floral changes in Asia have not been studied as well as those of North America or Europe. The most recent work was summarized by Leopold et al. (1992). A subtropical woody savanna developed in northwestern China during the late Eocene, lacking in evidence of grasses and herbaceous pollen (as were reported for other continents). By the late Eocene, subarid floras were found in northwestern China, while southeastern China was more mesic. Coastal areas had broadleaved forests, cypress swamps, and even mangroves. This northwest-southeast arid-humid gradient across China persisted into the early Oligocene, although with decreased diversity throughout the region. The woody savanna persisted into the late Oligocene, with increased importance of salt bushes. In highland and coastal areas, the increase in temperate deciduous trees and conifers suggest late Oligocene cooling. Forest vegetation with patches of woody savanna prevailed in Kazakhstan (once part of the former Soviet Union) in the early Oligocene.

Lost Worlds: Africa, Australia, South America

Charles Andrews's illness was getting worse in the cold, wet British winters. He worked at the British Museum of Natural History in London, where the labs were particularly damp and uncomfortable. His doctors urged him to spend his winters in some warmer, drier climate. Many of his friends went to the Mediterranean during the winter, or booked a long passage to one of the British colonies in the tropics, such as India. Andrews decided he would go to Egypt and collect some of the exotic living mammals for the Museum. After all, Egypt was a British crown colony, so he could be sure of safety and the support of local authorities. In April 1901, Andrews sailed to Alexandria.

When he arrived, he found plenty of mammals to trap or shoot for the collections. But something else caught his attention. Just a few years before, the British geologist H. J. L. Beadnell had been mapping the rocks at a barren, rocky area called the Fayum Depression in the Sahara Desert 70 miles southwest of Cairo (figure 5.13). Beadnell discovered abundant mammal bones lying on the desert floor, exposed by the howling winds. In just three weeks, Beadnell and Andrews made major collections of extinct mammals that included many large skulls, vertebrae, ribs, and teeth, as well as petrified wood. As Andrews collected, it became apparent that most of the fossils he found were unlike anything ever seen before. There were strange beasts with paired horns on their snouts, archaic whales, archaic mastodonts, and many other extinct forms that had no living descendants. Even more surprising was the lack of

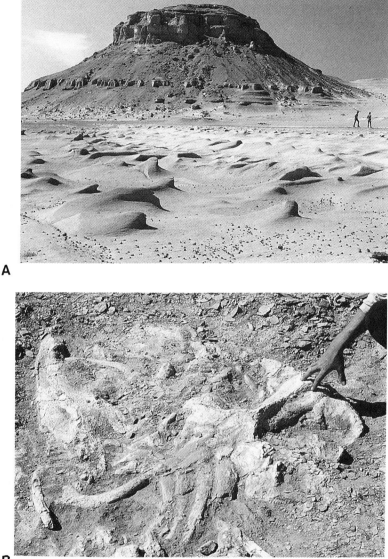

A

B

FIGURE 5.13. (A) Typical outcrops of the lower Oligocene Jebel Qatrani Formation of the Fayum region, Egypt, source of the best Oligocene fossils from Africa. (B) Partial skeleton of a fossil archaeocete whale as it appeared during excavations in the Fayum. (Photos courtesy D. Tab Rasmussen.)

animals typical of Eurasia: no carnivores, no rhinos or horses, almost no rodents, and none of the artiodactyls that were so abundant everywhere else. It was like stepping into a lost world.

Almost no fossil mammals had been known from anywhere in Africa until a German explorer of the Nile, Georg Schweinfurth, found fossils of a primitive whale in the Fayum in 1879. Beadnell and Andrews returned to the Fayum each winter for three years. By the time they finished in 1904, crates containing tons of fossil bones jammed the basement of the British Museum and the Egyptian Museum, waiting to be studied. Andrews had gone to Egypt for a rest, and came back with so much new material that his life's work was completely changed. Today he is best known for his descriptions of the bizarre beasts found on an Egyptian "vacation" (Andrews 1906).

The strangeness of the Fayum mammals, and the scarcity of mammals typical of Eurasia, clearly showed one thing. Africa must have been an island continent, mostly isolated from the Northern Hemisphere during the Paleocene, Eocene, and Oligocene. Some of the endemic groups of mammals never spread beyond Africa. Others evolved in Africa, but later escaped to Eurasia. African mastodonts, for example, spread around the world in the Miocene. Still others, the hyraxes (living marmot-like creatures that are distant relatives of perissodactyls and elephants) had radiated into ecological equivalents of pigs, hippos, cattle, antelopes, and horses. These animals filled the ecological niches in Africa of more familiar Eurasian ungulates, yet they evolved these shapes independently. Clearly, without the Eurasian mammals, there were unoccupied niches for pig-like or hippo-like forms to fill, no matter what their ancestry.

The Fayum desert and many other localities in Egypt were collected by a number of other scientists in the decades following the work of Beadnell and Andrews. German paleontologists Ernst Stromer von Reichenbach and Richard Markgraf collected there just before the First World War. They found the Fayum's most famous denizens: the first anthropoid primates, ancestors of apes and humans. In the 1960s, several expeditions from the Yale Peabody Museum renewed collecting in the Fayum, primarily because of the interest in our

FIGURE 5.14. The most spectacular beast of the Fayum, the rhino-size *Arsinoitherium*. Long considered a mystery among fossil mammals, arsinoitheres are now known to be distantly related to elephants and mastodonts. (Painting by Z. Burian.)

primate ancestors. The bulk of their collections is of large mammals from archaic African groups. These include the most primitive proboscideans (elephants, mastodonts, and their relatives), the aquatic sirenians (represented today by manatees and sea cows), the bizarre twin-horned *Arsinoitherium* (figure 5.14), as well as a variety of hyraxes and archaic whales. Because most of these animals first appear in Africa, it seems likely that they originated and diversified on that continent during its period of isolation.

The Fayum sequence is practically the only African record that spans the Eocene-Oligocene transition. In 1992 Tab Rasmussen, Tom Bown, and Elwyn Simons reviewed Fayum stratigraphy and mammal faunas. They found no significant faunal changes during the Oligocene deterioration—in contrast to what is known about faunas on other continents. However, Egypt was part of the tropical Tethyan belt during the early Oligocene. Although there was some cooling and extinction in the marine realm, tropical waters remained warm enough to buffer nearby land floras and faunas from significant climatic change. Consequently, North Africa was a refuge from the "big chill" that so severely affected temperate and polar latitudes.

South America was another "lost world," or island continent, isolated during the Eocene and Oligocene. Its native fauna consisted of endemic hoofed mammals (unrelated at the ordinal level to those found elsewhere in the world), carnivorous pouched mammals, and edentates (anteaters, armadillos, sloths, and their kin). As noted earlier, however, there is still a major gap in the South American land record between the Mustersan (middle or late Eocene, about 40 million years old) and the Divisaderan-Deseadan (late Oligocene, about 28 million years ago). The gap that spans the late Eocene through the mid-Oligocene also spans most of the time we are interested in, so no detailed assessment of the Eocene-Oligocene transition is possible at present. However, it is possible to compare some gross differences between middle Eocene and late Oligocene mammal faunas.

According to Marshall and Cifelli (1989), South America underwent a change in vegetation from subtropical woodlands to seasonally arid savanna woodlands between the Mustersan and Deseadan. This is reflected in the extinction of many archaic browsing mammals and their replacement by grazers with high-crowned teeth during the Deseadan (Cifelli 1985). Pascual et al. (1985) and Pascual and Ortiz Jaureguizar (1990) argue that the changes between the Mustersan and Deseadan were among the most fundamental in South American history. The transition marked the end of nearly all archaic groups, including not only the browsing ungulates, but also rodent-like marsupials, and primitive edentates. Many of these last occurrences represented the gigantic end-members of long-established lineages, such as the gigantic carnivorous marsupials. By contrast, lineages with smaller species, and the smaller, less specialized marsupial carnivores began to diversify in the

Deseadan. The most striking feature of the Deseadan is the immigration of the earliest New World monkeys and caviomorph rodents (including the living Guinea pigs, chinchillas, agoutis, capybaras, and their kin). Both of these groups are thought to have rafted from Africa during the Oligocene.

The final "lost world" or island continent is Australia. Until recently, there were no fossil mammals from Australian deposits older than late Oligocene, about 23 million years old (Tedford et al. 1975). In 1992 Godthelp and others reported the Tinga Marra local fauna, which is apparently early Eocene based on the similarity with South American Eocene marsupials. The surrounding clays have been dated at 54.6 ± 0.05 million years, consistent with that estimate. However, most of the fossils comprise a relatively poor and small sample (10 taxa based on 50 isolated teeth) of primitive marsupials, so it is premature to make statements about the diversity in the Australian Eocene. Whatever the eventual interpretation of this fauna, Australia still lacks fossil mammals for most of the Eocene and the early and mid-Oligocene. Neither Australia nor South America is helpful for studying the faunal changes at Eocene-Oligocene transition.

Kemp (1978) has reviewed the Australian botanical record. By the late Eocene, the rainforest vegetation began to decline in diversity. A major change took place during the Oligocene deterioration, when it appears that deepwater circulation between Antarctica and Australia commenced, leading to Antarctic glaciation. Oligocene plant assemblages in Australia are characterized by a low diversity of mostly cool-temperate plants tolerant of higher seasonality. The initiation of widespread brown coal swamps, however, suggests high rainfall in southeastern Australia, which would have been under the influence of rain-bearing westerly winds triggered by the circum-Antarctic current. Truswell and Harris (1982) showed that open forest habitats, with a more diverse herbaceous understory, began in the early Oligocene as a consequence of increased aridity. According to Christophel (1990), Oligocene floras of Australia show a great increase in stiff, fibrous leaves and a reduction of leaf size in many taxa, including conifers.

THE BIG CHILL

The Bigger Chill of the Mid-Oligocene

Although it triggered massive vegetational changes and significant extinctions and migrations in much of the world, the early Oligocene glacial event was not long-lived. According to Miller et al. (1991) and Zachos et al. (1992), the early Oligocene pulse of glaciation may have lasted only a few hundred thousand years, concentrated in western Antarctica and the Indian Ocean sector of the Southern Ocean. The glaciation apparently had limited effect in East Antarctica and the South Atlantic sector of the Southern Ocean (Kennett and Barker 1990). However, there is evidence in magnetic Chron C11N of a much bigger and more protracted glacial event or events starting about 30 milliion years ago in the late early Oligocene (sometimes called "middle" Oligocene, although the Oligocene has been formally divided only into early and late stages). Benthic foraminiferan $\delta^{18}O$ values exceeded 1.6‰ and eventually reached 3.0‰, which exceeds the threshold necessary for renewed glaciation (Miller et al. 1991; Miller 1992). These high glacial values of $\delta^{18}O$ persisted for about 4 million years, subjecting Antarctica to the longest glaciation (or multiple episodes of glaciation) it had ever experienced up to that point.

The effects of this great ice sheet are apparent all over the south polar region (figure 5.15). The CIROS-1 drill core in the Ross Sea penetrated almost 200 meters (650 feet) of mid-Oligocene glacial sediments (Barrett et al. 1989). Mid- or late Oligocene glacial sediments were also reported 200 km away in the MSST-1 core in Victoria Land (Barrett et al. 1987), on Marie Byrd Land (LeMasurier and Rex 1982), and in King Georges Island on the Antarctic Peninsula near South America (Birkenmajer 1987). Even more striking is the evidence from voyages of the research vessel *R/V Polar Duke* in the Ross Sea. By bouncing seismic waves off the layers in the sediments beneath the Ross ice shelf, a major mid-Oligocene unconformity was detected, overlain by hundreds of meters of upper Oligocene glacial sediments (Bartek and Anderson 1990; Bartek et al. 1992). Lou Bartek and colleagues were able to trace this unconformity up onto the Antarctic continent, and across at least 100,000 square kilometers beneath the Ross ice shelf.

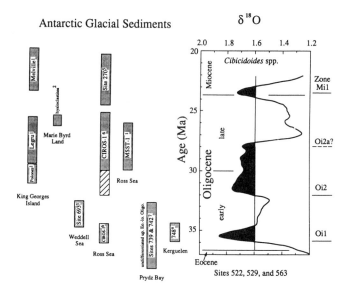

FIGURE 5.15. Evidence of Antarctic glaciation during the Oligocene comes from various glacial deposits (left) and from the strongly positive oxygen isotope values (right). Each time the oxygen isotope curve moves to the left past the 1.6‰ threshold, Miller et al. (1991) infer an episode of Antarctic glaciation (solid black areas). (By permission of the American Geophysical Union.)

Such a large feature could only have been produced by the grounding of a large ice sheet, comparable to the glacial activity in the Pleistocene.

The most striking global effect of this massive ice sheet was a worldwide drop in sea level. As sea level retreated, continental shelf would have been exposed to erosion. A major mid-Oligocene unconformity is indeed found in many marine shelf sequences around the world, including the U.S. Gulf Coast (Keller 1985), the Atlantic Coast of New Jersey (Olsson et al. 1980; Poag and Schlee 1984; Poag and Ward 1987), and the continental shelves of Europe (Aubry 1985), Australia (Quilty

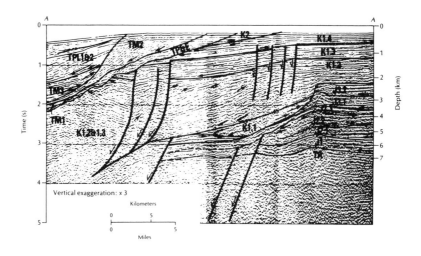

FIGURE 5.16. Typical seismic profile across the Atlantic margin of Africa, showing the major mid-Oligocene unconformity (between TP&E and TM1) (From Vail et al. 1977; by permission of the American Association of Petroleum Geologists.)

1977; Loutit and Kennett 1981; McGowran 1992) and southern Africa (Siesser and Dingle 1981). Deep-sea cores typically have a major gap representing the mid-Oligocene; some were so deeply eroded that the lower Oligocene was removed as well (Kennett et al. 1972; Keller et al. 1987). Seismic reflections from beneath the continental shelf off New Jersey and Virginia reveal deeply incised submarine canyons, cut by rivers when the retreating ocean left the shelf exposed in the mid-Oligocene (Miller et al. 1985, 1987).

Seismic data of this sort led to one of the most exciting and controversial discoveries of the last two decades: the "Vail sea level curves." A group led by Peter Vail at Exxon accumulated seismic profiles across the passive margins on both sides of the Atlantic, from North America to Brazil, and from Africa to Europe (figure 5.16). The Vail group traced seismic reflectors in the sections through the continental shelves, and found that

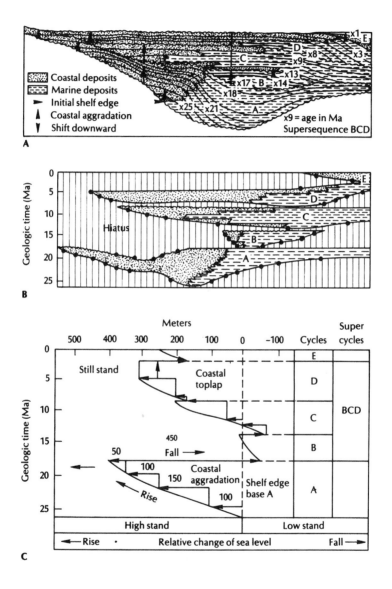

FIGURE 5.17. Procedure for interpreting onlap-offlap history from seismic profiles (From Vail et al. 1977; by permission of the American Association of Petroleum Geologists.)

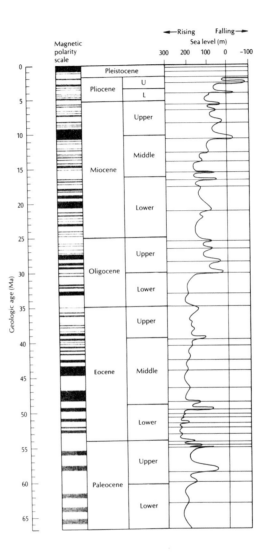

FIGURE 5.18. "Vail curve" of sea level changes, showing the major drop in the mid-Oligocene, about 30 Ma. Note that this time scale has not been updated to reflect the new dates, and so it places the Eocene/Oligocene boundary at 35, not 34 Ma. (Modifed from Haq et al. 1987.)

they could subdivide the entire passive margin sequence into discrete packages of sediment bounded by unconformities. Each package represented onlap of the marine sediments when sea level rose across an erosional surface, and was truncated by erosion due to offlap when sea level fell (figure 5.17). By tracing out these unconformity-bounded sequences and noting how far each unconformity extended toward land, it was possible to reconstruct the history of onlap and offlap since the Jurassic. In 1977, Vail and his colleagues published their results. The "Vail sea level curve" was the first detailed record of the last 150 million years of global sea level history (figure 5.18).

Soon thereafter, scientists began to realize that the Vail curve was not strictly controlled by sea level. For example, the Vail curve showed regressions happening more rapidly than transgressions, giving the curve a "sawtoothed" shape with smooth, slow trangressions and flat-topped, rapid regressions (figure 5.17C). During regression the nearshore and non-marine sediments can prograde out onto the shelf, building up the sedimentary package even when sea level is falling. Thus most people now refer to these cycles as the "Vail onlap-offlap" curves, and the updated versions (Haq et al. 1987, 1988) show a smooth, symmetrical sea level curve alongside the sawtoothed onlap-offlap curve.

Vail and colleagues have been criticized (summarized by Hallam 1992) for claiming that their curve represents the history of global sea level. Unfortunately, the data are unavailable for study, because they are property of Exxon. Outside parties cannot analyze the basis for the claims of the Vail group. In some areas, local or regional tectonic events may be more responsible for unconformities (Cloetingh et al. 1985; Hubbard 1988; Hallam 1988, 1992). Vail's data set may also be flawed because it comes almost exclusively from the Atlantic margin, which has a very uniform subsidence history as it was all formed by the same process of sea-floor spreading (Hubbard 1988). In support of the Vail hypothesis, stratigraphic studies of exposed marine sequences have shown that most of the unconformities of Vail et al. (1977) and Haq et al. (1987, 1988) can be traced to surface outcrops (Wilgus et al. 1988; Poag and

Ward 1987). However, Miall (1992) points out that the Haq et al. (1987, 1988) curve has so many "events" that almost any local sea level record can be made to "fit" the curve. Thus, local tests of the "global" record may not prove much.

Of all the unconformities in the Vail curve, the most striking is the one in the mid-Oligocene (figure 5.18). According to Haq et al. (1987, 1988), sea level dropped by over 150 meters (over 500 feet), greater than any regression during the Pleistocene or any other time in the last 150 million years. Some have suggested that this figure is a bit exaggerated (Loutit and Kennett 1981), but the widespread mid-Oligocene unconformity and the huge submarine canyons described above show that it was undisputably a very big event. Huge Antarctic ice sheets must have locked up an enormous amount of water in the Ross Sea region, and continental shelves were exposed all over the world.

Mid-Oligocene Survivors

Such a major cooling event ought to have caused extinctions greater than those at the end of the middle Eocene or earliest Oligocene. The mid-Oligocene cooling was accompanied by diversification of cold-water planktonic foraminiferans (Keller 1983a, 1983b) and coccoliths (Haq and Lohmann 1976; Haq et al. 1977; Aubry 1992), but there were relatively few extinctions. No significant extinctions affected the diatoms (Baldauf 1992). Although the data are sparse, few changes affected the molluscs (Hansen 1987, 1992) or echinoids (McKinney et al. 1992).

If the mid-Oligocene cooling was so severe, why did it not cause even more severe extinctions? A number of scientists have suggested that the mid-Oligocene biota were less susceptible to extinctions. Most of the organisms that became extinct at the end of the middle Eocene were tropical taxa, specialized for a narrow range of temperatures and other conditions. The late Eocene and early Oligocene were times of repeated episodes of climatic deterioration, but recurrent extinctions proved less and less severe as the vulnerable tropical specialists were eliminated. In the planktonic foraminiferans, for example, Gerta Keller (1983a, 1983b) has shown that the only species left by the mid-Oligocene were hardy forms tolerant of cold water, which would not be adversely affected by another cooling episode (figure 5.3). Steve Stanley (1990) has suggested that this phenomenon is true of almost all the major mass extinctions. Such crises tend to wipe out the specialists (concentrated in the tropics), leaving a low diversity of hardy, cold-tolerant generalists. In the five or ten million years following a mass extinction, these generalists gradually diversify and a new wave of tropical specialists appears. If another crisis hits in the first few million years after a mass extinction, it might not have much effect on the biota, because few vulnerable taxa would yet have evolved. Finally, fifteen to twenty million years after the mass extinction, the earth recovers a diversity comparable to before the previous extinction; the biota again becomes vulnerable.

In the case of the Cenozoic, the K/T extinctions left only a few hardy generalists through much of the Paleocene. The global diversity of organisms reached a peak in the early and early middle Eocene, which coincided with a peak of warming. Then a series of shocks at the end of the middle Eocene and in the earliest Oligocene left only hardy survivors to experience the subsequent great glaciation and global cooling of the mid-Oligocene.

Similar patterns can be seen on land. Wolfe (1978) detected no significant mid-Oligocene effect on North American land plants after the earliest Oligocene refrigeration, probably because late early Oligocene floras were already cold adapted, and tropical taxa had been driven to extinction or to lower latitudes (figure 3.7). In the Badlands paleosols, Retallack (1983a, 1983b, 1990, 1992) described a transition from early Oligocene (Orellan) wooded grasslands and gallery woodland to late Oligocene (Whitneyan and early Arikareean) open grasslands with trees only along watercourses (figure 5.7B). There is also a drying trend, from average annual rainfall of 500–900 mm (20–35 inches) in the early Oligocene to 350–450 mm (14–18 inches) in the late Oligocene. Whitneyan and early Arikareean deposits in the Great Plains were predominantly wind-blown silts and volcanic dust, indicating very dry conditions (figure 5.19).

FIGURE 5.19. Typical outcrops of the Whitneyan Poleslide Member of the Brule Formation, from the top of Sheep Mountain Table, Badlands National Park, South Dakota. These wind-blown siltstones were formed during drier conditions, and the resistant ledges of sandstone were deposited in seasonal river channels. In the old terminology (figure 2.11), the Whitneyan was called the "*Leptauchenia* beds" (see figure 5.20) and the river deposits were known as the "*Protoceras* channels" (see figure 5.21A) due to the abundance of these fossil mammals.

North American mammal faunas reached an all-time low in diversity in the Whitneyan (Prothero 1985; Stucky 1990, 1992). A little oreodont known as *Leptauchenia*, which had high-crowned teeth for grazing tough, gritty grasses was the most abundant mammal. Its eyes, ears, and nostrils were on the top of its head (figure 5.20). Some have interpreted this as an

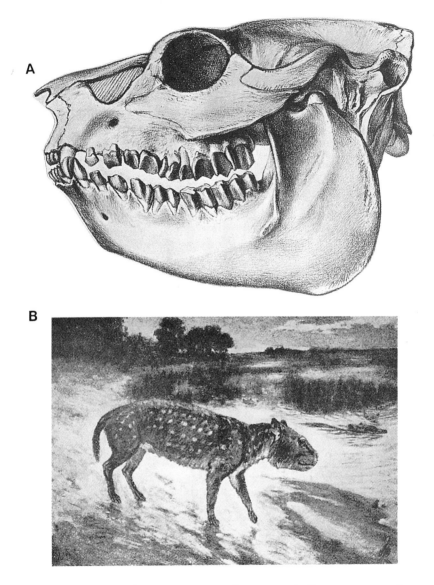

FIGURE 5.20. The common Whitneyan-Arikareean oreodont *Leptauchenia*. The skull (A) had very high-crowned teeth and eyes and nostrils located high on the head. (From Scott et al. 1941). (B) Restoration of *Leptauchenia* as an amphibious swimmer in an attempt to explain the eyes and nostrils on top of the head. However, their abudance in undoubted sand dune deposits makes this restoration doubtful. (From Scott 1913.)

215

aquatic adaptation, but these creature were found in great abundance in the arid dunes of volcanic dust that dominated the High Plains in the late Oligocene. The high-crowned teeth suggest a much grittier, coarser diet. Moreover, there are living desert animals with high-placed nostrils and eyes, and openings on the snout like *Leptauchenia* for the filters that keep dust out of the lungs. A dwarfed leptauchenid known as *Sespia* was particularly common in the early Arikareean, especially in California.

The more generalized oreodonts, the merycoidodonts, became much larger, reaching almost cow-sized proportions. Three-toed horses, such as *Miohippus*, also got larger than their Orellan forbears, as did the true rhinoceros with paired horns on its snout, *Diceratherium*. The archaic amynodont and hyracodont rhinos were both extinct by the early Arikareean. Instead, stream-side forests were occupied by antelope-like relatives of camels known as *Protoceras* (figure 5.21A); their fossils show a series of knobs and tusks on the skulls of males (presumed females were hornless). The Whitneyan was informally known as the "*Protoceras* beds," although *Protoceras* is found only in the stream channel facies, and only from South Dakota (figure 5.19). With their very high-crowned teeth and long necks and legs, camels also began to diversify in the Arikareean.

In addition to oreodonts and camels, two other groups of mammals diversified in the drier, harsher conditions of the late Oligocene in North America. The Canidae (dog family) underwent an adaptive radiation in the early Arikareean, taking a greater variety of body sizes and feeding specializations. Even more impressive is the change in rodent faunas. Many of the archaic rodent groups from the Eocene and early Oligocene were nearly gone, and in their place were several modern rodent families with higher-crowned teeth and other adaptations for feeding and living in dry, grassy conditions. The most abundant were the cricetids (now represented by the hamster and New World mice), but pocket mice, jumping mice, and pocket gophers were also common. The abundance of pocket gophers, and the appearance of the horned burrowing rodents known as mylagaulids, suggests that burrowing became

216

A

B

FIGURE 5.21. (A) The multihorned Whitneyan artiodactyl *Protoceras*. Distantly related to camels, the protoceratids evolved into deer-like creatures with a variety of odd horns, including some with a bony "slingshot" horn on their skull (figure 5.22). (B) Early Arikareean deposits in western Nebraska include remarkable filled burrows known as "Devil's corkscrews" or *Daemonelix*. They are now known to have been dug by *Palaeocastor*, a burrowing relative of the beaver. (From Scott 1913.)

217

important in surviving the drier conditions (as we see in burrowing rodents of grasslands and deserts today). Filled casts of the corkscrew-shaped burrows (called *Daemonelix*, or "Devil's corkscrews") of the burrowing beaver *Palaeocastor* are among the most distinctive fossils in the Arikaree Group of Nebraska (figure 5.21B).

Faunas of the Asian mid-Oligocene ("late Oligocene" of Li and Ting 1983, Russell and Zhai 1987, and Wang 1992) had a few similarities to North American faunas, although the exchange across the Bering Strait had shut down in the early Oligocene. As in North America, the apparent diversity of Asian mammals was very low, although the taxonomy has not been fully updated. The dominant rodents were again cricetids and jumping mice, along with the ctenodactylids (gundis) and rhizomyids (bamboo rats and mole rats). Rabbits abounded. The gigantic indricotheres still persisted, and they would survive until about 15 million years ago in Pakistan. True rhinoceroses were also characteristic of the Asian mid-Oligocene, as were the clawed chalicotheres, the hippo-like anthracotheres, and a variety of small deer-like ruminants. The last of the pig-like entelodonts remained, but the first true pigs (Family Suidae) lived side-by-side with them. The archaic creodont *Hyaenodon* had not yet disappeared, but bears, raccoons, beardogs, and cat-like nimravids were the dominant carnivorans.

European faunas were also low in diversity in the middle and late Oligocene. There is a slight decline in diversity, which seems to indicate further cooling (Vianey-Liaud 1991). Some-time in the mid-Oligocene (between the Sannoisian and Stampian "stages"), a few palaeotheres, anthracotheres, and other archaic genera suffered extinction, but some modern groups (such as true pigs, and advanced ruminants) replaced them (Brunet 1977; McKenna 1983a). Typical late Oligocene faunas were characterized by abundant ruminant artiodactyls and deer, although archaic Eocene groups like anoplotheres, cainotheres, and even the truly primitive dichobunids survived. True rhinoceroses were the dominant perissodactyls, but one of the last of the amynodonts, a tapir-like beast with a proboscis known as *Cadurcotherium*, was hanging on. Although the creodont *Hyaenodon* was still abundant, carnivorans in the modern families that include bears, raccoons, weasels, mongeese, plus the first true cats, were the dominant predators. Eocene rodent families like the Theridomyidae and Eomyidae were still common, but aplodontids, squirrels, beavers, dormice, and especially cricetids were the dominant rodents. At the threshold of the Miocene, European mammals had become recognizably modern.

Miocene Recovery

Although the Eocene and Oligocene are the focus of this book, it is important to gain a sense of what followed. Based on oxygen isotopic records and Antarctic glacial deposits (Miller et al. 1991), the mid-Oligocene cooling and glaciation seems to have ceased by the latest Oligocene (figure 5.15), about 26 million years ago. One more brief pulse of glaciation marks the Oligocene-Miocene boundary (about 24 million years ago), but the rest of the early Miocene shows a worldwide warming trend (Miller et al. 1987; Shackleton 1986). Marine organisms from plankton to molluscs to whales recovered and diversified, so that by the time of the mid-Miocene resumption of Antarctic glaciation, there were many warm-water specialists vulnerable to extinction. The complex history of glaciation, oceanic circulation changes, and extinctions during the 19 million years of the Miocene is too complex to discuss here, but there are many published studies (summarized in Kennett 1985).

The Miocene in North America shows a thermal and floral recovery from the Oligocene chill (Wolfe 1978, 1981). There was, however, a general drying trend that favored the development of grassy savannas (Webb 1977). Land mammal diversity also recovered (Stucky 1990, 1992), but the mammals were predominantly grassland specialists. Horses, rhinos, camels, and ruminants all showed a tendency toward longer limbs for escaping predators in a landscape where speed mattered, and higher-crowned teeth adapted for grazing (Webb 1977, 1983; Tedford et al. 1987). Similar trends can be documented for Europe and Asia (Savage and Russell 1983), although the data are not as detailed and well-calibrated as they are for North America.

FIGURE 5.22. Mammals of the North American Miocene savanna included gazelle-like and giraffe-like camels (right), many types of horses (center), mastodonts (extreme right), a slingshot-horned protoceratid (foreground), and giant pig-like entelodonts (left). (Courtesy Yale Peabody Museum.)

By the mid-Miocene (about 15 million years ago), most of the northern continents had fully developed savanna vegetation (figure 5.22). In North America, for example, there are Miocene fossil localities all over the west, and Florida is full of mammal bones. These mammal communities had many similarities to the modern East African savanna, only the cast of characters was slightly different (Webb 1983; Prothero and Schoch 1993). As many as 12 different species of three-toed horses lived side-by-side, some adapted for exclusive grazing and others retaining low-crowned teeth for browsing. Some rhinoceroses browsed on bushes (like the modern black rhino), but *Teleoceras* developed into a fat, hippo-like beast that lived in the water, presumably coming out at night to graze. Anatomically, *Teleoceras* was a rhinoceros, but it was the ecological equivalent of hippos.

Camels were North American natives who remained confined to North America until the Pleistocene, when they escaped to the Old World (evolving into dromedaries and Bactrians) and South America (resulting in llamas, vicuñas, and guanacos). In the North American Miocene they were extremely diverse, evolving into gazelle-like stenomylines and gigantic "giraffe-camels" whose leg and neck proportions mimicked today's giraffes. Pronghorns were also diverse, and sported a variety of horn shapes, like modern African antelopes. But pronghorns then and now are not true antelopes, which are related to cattle and goats; they are North American natives related to deer. The native oreodonts, which had been abundant in the Oligocene, became more specialized in the Miocene. Some were pig-like, and others had a long, tapir-like proboscis. Also native to North America were the pig-like peccaries. They filled the pig niche in North America, while true pigs were a strictly Old World group.

Instead of deer, North America harbored a distant relative, the dromomerycids, which displayed a variety of bony horns on their snouts, over their eyes, and on the tops of their heads. There were blastomerycids, small and hornless deer-like ruminants with tusks, very similar to the musk deer now found in Asia. The strange multi-horned *Protoceras* (distantly related to camels) of the Whitneyan (figure 5.21A) was succeeded by a variety of bizarre protoceratids in the Miocene. Some had bony Y-shaped "slingshots" on their snouts, and others had long and curved bony horns over their eyes.

As on the African savanna, the largest herbivores in North America were elephant relatives, the mastodonts. They ended their isolation in Africa during the early Miocene when the Middle Eastern connection closed the Tethys seaway. From there, they spread around the northern hemisphere, reaching North America about 16 million years ago. The strangest mastodonts evolved huge flat jaws that looked like snow shovels; others had four long straight tusks.

North American Miocene predators included not only members of the dog, cat, bear, weasel, and raccoon families, but

221

also extinct carnivorans such as the amphicyonids, or "bear-dogs," several species of saber-toothed cat unrelated to the Ice Age sabertooth *Smilodon*, and specialized hyaena-like dogs—the borophagines—with teeth and jaws adapted for bone-crushing. Abundant rodents lived in the grasses or trees, including squirrels, horned mylagaulids, beavers, pocket gophers, pocket mice, cricetids, and jumping mice, plus a variety of rabbits.

Animal for animal, then, the parallels between North America in the Miocene and the modern East African savanna are striking. Evidence of similar Miocene savannas has been found in Europe, Pakistan, and China, with different families of mammals performing the same ecological roles. Even South America, which had mostly native hoofed mammals unrelated to those in the rest of the world, fit this pattern. Some Miocene hoofed mammals of South America evolved extraordinary parallels with mastodonts, hippos, rhinos, camels, giraffes, antelopes, and even horses. The South American "horse" equivalent not only developed high-crowned teeth and long limbs, but it completely lost its side toes—it was even more one-toed and horse-like than the living horse.

Thus, a kind of savanna "Garden of Eden" characterized much of the world during the warmer but drier climate of the early and middle Miocene. As glaciation and cooling resumed in the middle Miocene, this habitat and all its denizens were eventually wiped out (except in equatorial East Africa). But that story is beyond the scope of this book. Instead, the final chapter examines the possible causes of our main subject: the great changes of climate during the Eocene and Oligocene.

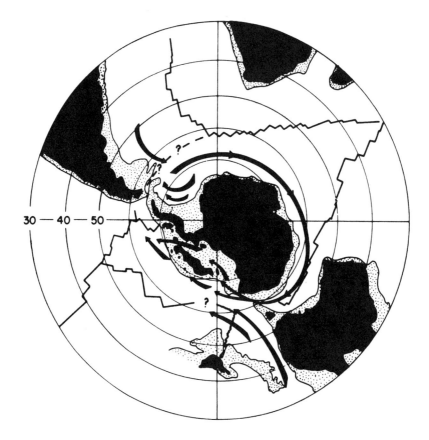

FIGURE 6.1. Oligocene circulation patterns, showing the beginning of the circum-Antarctic current as Australia separated from Antarctica. (Modified from Kennett 1982; by permission of Prentice-Hall, Inc.)

6

Glaciers, Volcanoes, or Asteroids?

False facts are highly injurious to the progress of
science, for they often endure long; but false views,
if supported by some evidence, do little harm,
for every one takes delight in proving their falseness.

—CHARLES DARWIN, *THE DESCENT OF MAN* (1871)

For every complex problem there is a solution
that is simple, neat, and wrong.

—H. L. MENCKEN

Previous chapters have reviewed the evidence of the cooling and ice-volume changes that occurred sporadically in the middle and late Eocene and throughout the Oligocene. Changes in the temperature tolerances of marine microfossils and land plants indicated periods of cooling and drying, and even glaciation. This chapter examines explanations for the end of the Eocene greenhouse.

Not with a Bang but a Whimper
Chapter 4 demonstrated how the scientific bandwagon of asteroid-induced mass extinctions soon focused on the Eocene. At least four separate tektite and/or iridium horizons (Miller et al. 1991b) are dated between 35.5 and 36.0 million years ago, in magnetic Chrons C15N and C16N. This puts them about 1 to 2 million years before the Eocene-Oligocene

boundary (which was not a major extinction event or climatic horizon) and almost 3 million years before the early Oligocene refrigeration event. These tektite/iridium horizons are about 2 million years too late to account for the widespread extinctions of the middle Eocene. These impacts during the middle of the late Eocene were at the wrong time to have any direct bearing on mass extinctions.

The problem of timing points out a general problem with attempts to explain extinctions with extraterrestrial impacts: loose coincidence of extraterrestrial events and extinctions does not prove a direct causal relationship. For the impact scenario to be convincing, there must be tight correlation in time between the impact cause and the extinction effects, and the pattern of extinctions should make sense in terms of likely effects on global climate. Instead, we find strong evidence of major global climatic changes starting in the middle Eocene, well before any extraterrestrial objects arrived, and most extinctions seem tightly correlated with these climatic shifts. Glass (1982) and Hut et al. (1987) raised the possibility that the impacts may have triggered climatic change, but the impacts were millions of years too late for the middle Eocene events, and too early for the Oligocene events.

When impacts occur, their effects should be felt within a few minutes to a few thousand years. There are no known earthly processes that would allow extraterrestial perturbations to persist over millions of years, particularly if the impacts themselves seem to have had little effect in the immediate aftermath. The world of the latest Eocene (after the impacts) shows no evidence of disturbance by forces from space. Rather, the climatic shifts initiated much earlier and much later include major changes in oceanic circulation and growth of ice sheets; these must require more sustained forces to develop. How could ephemeral impacts explain the spread and steady production of deep, cold Antarctic bottom waters? Finally, the pattern of extinction (from tropical marine organisms to changes in land plants) seems to be strongly temperature-related. When we see isotopic evidence of changes in oceanic temperature and circulation, there were also responses in the marine organisms.

FIGURE 6.2. The eruption of Krakatau volcano, west of Java, as portrayed in a contemporary book. (Courtesy U.S. Geological Survey.)

If impacts were unimportant in the overall scheme of Eocene climatic change, what about volcanism? Since the explosion of Krakatau in 1888 (figure 6.2), we have known that volcanism can cause cooling, and volcanic activity often persists for millions of years. Indeed, volcanism has been offered as an explanation for the K/T extinctions (Officer et al. 1987; Courtillot 1990), and for the greatest extinctions in all earth history, the Permo-

Triassic (Erwin 1993). Kennett et al. (1985) reported many volcanic ash layers in deep-sea cores taken in the southwest Pacific, mostly due to intense volcanic activity in New Zealand and along the boundary between the Pacific and Indo-Australian plates. However, most of these layers are concentrated in the latest Eocene and early Oligocene, too late for the major extinctions and cooling at the end of the middle Eocene. The volcanism was also too protracted to have caused the sudden early Oligocene refrigeration, although it may have served as a trigger.

Rampino and Stothers (1988) have argued that mass extinctions are correlated with major floods of basaltic lavas, erupting from great fissures in the crust that brought up magma from the mantle. Flood basalts have erupted in different parts of the world throughout earth history, and some scientists have claimed that their eruptions show the same periodicity alleged by Raup and Sepkoski (1984). Rampino and Stothers attributed both the cause and the alleged periodicity of these flood basalts to a hypothetical periodicity in comet impacts. Others (Loper and McCartney 1986; Loper et al. 1988) have suggested that the mantle undergoes periodic pulses of upwelling, causing periodicity in flood basalts. Whatever the ultimate cause, Rampino and Stothers suggest that the great Eocene eruptions of basalt in Ethiopia, which cover over 750,000 square kilometers (Berhe et al. 1987), might be related to late Eocene extinctions.

Over 122 dates have been run on these Ethiopian basalts, giving a spread of ages between 10 and 80 million years ago. The earliest massive eruptions were in the early Eocene, about 49 million years ago—too early to have contributed to the middle Eocene extinctions (Berhe et al. 1987). The peak of eruptions seems to have occurred between 35 and 25 million years ago, and Rampino and Stothers (1988) estimate that the initiation of major eruptions started around 35 ± 2 million years ago. Once again, this is too late for the middle Eocene extinctions. More important, like the southwest Pacific eruptions, the Ethiopian basalt ages are too spread out, and do not show a clear peak in activity that could be considered a proximal cause of extinction.

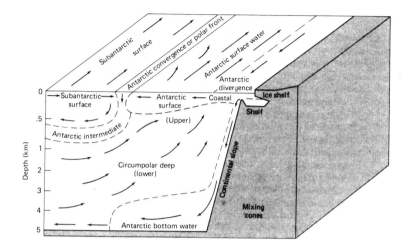

FIGURE 6.3. Diagram showing the generation of water masses around Antarctica, including the shallow circum-Antarctic current and the deep Antarctic bottom water. (Modified from Kennett 1982; by permission of Prentice-Hall, Inc..)

The Search for a Cause

The volcanism scenario falters before the same problems as the impact hypothesis: it cannot explain the long-term changes in oceanic circulation that clearly controlled the extinctions of tropical life. Instead, we must look to protracted terrestrial processes that can dramatically shift ocean currents and water masses. These control the formation of Antarctic ice and ultimately temperature and precipitation everywhere on earth. The only known process that can produce long-term changes in oceanic circulation is plate tectonics.

Chapter 3 presented the record of isotopic ratios and planktonic foraminiferans that implicated the Antarctic as a major source of cooling in the Eocene. When the first clear isotopic and micropaleontological data began to emerge from the Southern Hemisphere (Devereux 1967), several authors

229

(Kennett et al. 1972, 1975; Frakes and Kemp 1972) suggested that the critical factor was the separation of Antarctica and Australia. Today, the two continents are separated by a deep-water passage hundreds of miles wide, which allows cold Antarctic surface waters to flow easterly and circle completely around the Antarctic continent (figure 6.1). As the shallow Antarctic Circumpolar Current (ACC) cycles, it causes upwelling of circumpolar deep water (figure 6.3). The ACC is one of the most voluminous of ocean currents, with a velocity of about 25 cm/sec. The volume that passes between Antarctica and Australia is about 233 million cubic meters per second, or more than 1000 times the flow of the Amazon River (Callahan 1971)!

The induced upwelling brings up deep nutrients that support the enormous biogenic productivity of Antarctic waters, from phenomenal blooms of diatoms all the way to the whales at the top of the food chain. As the circumpolar deep waters reach the surface, they are "aired out" before they sink again. In addition, the Southern Ocean connects the Pacific, Atlantic, and Indian Oceans, moderating differences between them.

The ACC locks in the cold air and water around the South Pole and inhibits an exchange northward with warmer currents (figure 6.4). As circumpolar deep waters well up, they then chill and sink again to form deep, cold bottom waters which flow northward, creeping along the bottom of the Atlantic and Pacific all the way to high northern latitudes. These cold Antarctic bottom waters (AABW) make up 59% of the world's marine water (Warren 1971), and bring cold, oxygenated water as far north as 50°N in the Pacific and 45°N in the Atlantic. Such voluminous flows dominate the circulation patterns of the modern world ocean.

In the early Eocene, both Australia and South America were still attached to Antarctica as remnants of Gondwanaland. Africa and India had already detached. Instead of circumpolar circulation, the waters of the South Pacific and South Atlantic circulated with equatorial waters (figure 1.12). By doing so, they transported heat between poles and equator, and moderated the temperature extremes between high and low latitudes. This may partly explain why the Arctic and Antarctic were so balmy

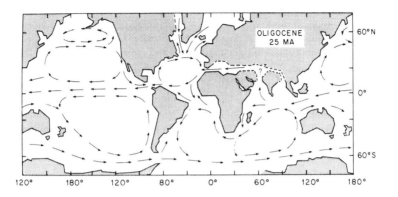

FIGURE 6.4. Oceanic circulation during the Oligocene. The circum-Antarctic current separates the tropical oceans from the Antarctic waters, preventing the level of mixing seen in the Eocene (compare with figure 1.12). Note also that India has nearly collided with Asia, restricting the flow through the tropical Tethys. (Modified from Frakes 1979.)

in the early Eocene, although calculations have shown that oceanic circulation alone is not sufficient to explain all the warmth (Barron 1985). Given the striking differences between the Eocene and modern circulation and climatic patterns, it is obvious that the development of the Antarctic circulation pattern is one of the major factors inducing Eocene climate changes and extinctions. The real argument centers around whether it is the only critical factor, and when its effects began.

When the evidence for the beginning of circum-Antarctic circulation first emerged, attention focused on the timing of the separation of Antarctica and Australia. The two continents began to separate in the late Cretaceous and Paleocene (Weissel et al. 1977; Mutter et al. 1985; Veevers 1986), with rifting opening from west to east (McGowran 1973; Kennett et al. 1975). By the middle and early late Eocene there was oceanic sea floor spreading between the continents. Micro-fossils suggest that there was still only a shallow marine gulf

231

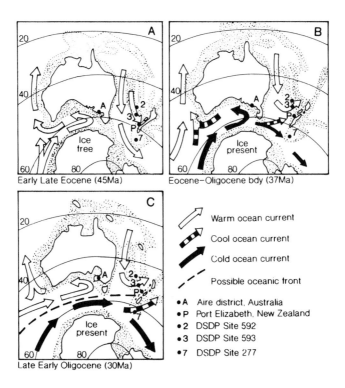

FIGURE 6.5. Opening of the shallow and deep-water passage between Tasmania and Australia during the later Eocene and Oligocene, showing reconstructed circulation patterns. (A) In the late Eocene, the shallow gulf between Antarctica and Australia provided moisture for the earliest Antarctic glaciers. (B) In the early Oligocene, the first deep-water flow (dark arrows) occurred through the passage south of Tasmania. (C) By the mid-Oligocene, both shallow (open arrows) and deep currents could pass through the strait. This initiated a separation of currents and the beginning of the boundary, or oceanic front (dashed line), between them. (Modified from Kamp et al. 1990.)

between the continents (figure 6.5A), with deep water flow apparently blocked from the South Pacific by the South Tasman Rise (Kennett et al. 1975; Kennett 1977, 1978, 1980; Murphy and Kennett 1986; Kamp et al. 1990). In 1973 Leg 29 of the

Deep Sea Drilling Project drilled in the South Pacific and discovered the first evidence of shallow marine circulation over the Tasman Rise; it apparently began in the latest Eocene (Shackleton and Kennett 1975; Kennett et al. 1975). This explains why Antarctica began to cool and why deep bottom waters were formed shortly thereafter in the earliest Oligocene (figure 6.5B).

Continued spreading between Antarctica and Australia enlarged the seaway between them. The sudden expansion of glacial ice and the widespread "middle" Oligocene unconformity (see chapter 5) suggests that a new phase of circumpolar circulation had begun by the end of the early Oligocene. Based on the isotopic and micropaleontological evidence, several authors (Kennett et al. 1975; Kennett 1977, 1978, 1980; Murphy and Kennett 1986; Kamp et al. 1990) have attributed this massive cooling event to the beginning of deep-water circulation through the gap between the South Tasman Rise and Antarctica. This is particularly apparent in DSDP Site 277, which lies on the submarine Campbell Plateau (south of New Zealand); it is right in the path of currents flowing through the gap. In the late Eocene, Site 277 was still bathed in warm waters flowing down the coast of eastern Australia, but as time passed, the differences in isotopic values between shallow and deep-dwelling foraminiferans increased. Just before the major unconformity of mid-Oligocene age, a sharp increase appears in the difference in isotopic values between Site 277 and sites northwest of New Zealand (still bathed in warm currents). According to Murphy and Kennett (1986) and Kamp et al. (1990), this is evidence that a blast of deep, cold water was passing south of Tasmania, separating the Antarctic circumpolar current from currents flowing further north (figure 6.5C). Once these currents were decoupled, the modern Antarctic polar front, separating the circum-Antarctic current from the more northerly currents, was established (figure 6.4).

Given the scale of the mid-Oligocene glaciation event and sea-level drop, it seems that circum-Antarctic circulation must also have occurred between Antarctica and South America through the Drake Passage (figure 6.1). Studies in this region (Barker and Burrell 1977, 1982; Sclater et al. 1986) suggest that

233

the Drake Passage did not open until the late Oligocene. However, the timing on this is not as well constrained in terms of biostratigraphy or magnetics. Thus, the opening of the Drake Passage might explain either the long duration of the mid-Oligocene cooling event (possibly event Oi2a of Miller et al. 1991) or the renewed cooling and glaciation at the Oligocene/Miocene boundary (event Mi1 of Miller et al. 1991). If Barker and Burrell (1982) are correct, circulation through the Drake Passage did not really commence until the Oligocene/Miocene boundary (figure 5.15).

Although the Southern Ocean seems to have been the major "cold spigot" providing deep, cold bottom waters, it was not the only source. A major unconformity of the early Oligocene in the North Atlantic produces reflection horizons in the seismic records of the Atlantic margin (Miller and Tucholke 1983; Mountain and Tucholke 1985). In addition, studies of benthic foraminiferans show that in the early Oligocene, the North Atlantic bottom waters were richer in ^{13}C than those of the Pacific (Miller and Tucholke 1983; Miller and Fairbanks 1983; Miller and Thomas 1985; Miller 1992). This suggests that some of the deep waters of the Atlantic must have come from the north. The Arctic Ocean had been isolated from the rest of the world's oceans since the Mesozoic, developing its own cold waters with high ^{13}C ratios. When the Arctic was reconnected with the North Atlantic, such cold waters would be analogous to the modern North Atlantic Deep Water (NADW). The best candidate for this marine passageway was the Norwegian-Greenland Sea, which apparently opened in the early Oligocene (Talwani and Eldholm 1977; Berggren 1982). Other possibilities include the Faeroe-Shetland channel and the Denmark Straits (Miller and Curry 1982).

There is no shortage of potential oceanographic sources for the cooling and climatic changes in the Oligocene. The newly-established pattern of shallow-water circum-Antarctic circulation plus the cold Arctic source in the North Atlantic probably triggered the earliest Oligocene cooling event that produced the first significant Antarctic ice sheets. The initiation of a deep-water passage south of Tasmania in the middle Oligocene further accentuated the cold trend, producing a major Antarctic ice cap and the world's largest sea level drop of all time. The completion of circum-Antarctic circulation through the Drake Passage accelerated the refrigeration, whether it happened in the late middle Oligocene, or at the Oligocene-Miocene transition.

The "Doubt House" World

The greenhouse world of the early Eocene thus tranformed in the early Oligocene into the "icehouse" world we still inhabit. There is clear evidence of Oligocene Antarctic ice sheets, and reasonable oceanographic-tectonic explanations for why this cooling developed. However, the isotopic evidence of a middle and late Eocene cooling, and isolated mountain glaciers, cries out for explanation.

There is evidence of middle Eocene mountain glaciers on the Antarctic Peninsula, although local botanical evidence suggests that the climate over most of Antarctica was cool temperate. By the late Eocene, evidence of icebergs appears on the Indian Ocean side of Antarctica, although there is none on the Atlantic side. Clearly, the global cooling had commenced, and Antarctica was beginning to experience climatic deterioration, even if it was still covered by cool temperate forests and isolated mountain glaciers, rather than a continental ice sheet. If there was no circum-Antarctic current until the latest Eocene and early Oligocene, then why did global cooling commence in the early part of the middle Eocene? Between the early Eocene "greenhouse" and the Oligocene "icehouse" was the middle and late Eocene "doubt house" (Watkins and Mountain 1990; Miller et al. 1991), which has no clear modern analogues or simple explanations. From about 50 until 34 million years ago, the planet was in a transitional phase that is only now beginning to be studied and understood.

Although there may not have been a deep marine seaway between Antarctica and Australia until the latest Eocene, at least a shallow marine gulf south of Australia was exposed to Indian Ocean circulation since rifting had begun in the middle Eocene. The most rapid phase of opening occurred during Chrons C18R–C19N, which is slightly before the middle-late Eocene boundary (Weissel et al. 1977; Mutter et al. 1985). In

235

Australia, this sudden increase in spreading rates from about 2 cm/yr to 5.5 cm/yr caused a major transgression of the seas onto the continent, the Khirthar transgression (McGowran 1978, 1990). Other regions, such as the U.S. Gulf Coast (Hansen 1987), also show a Chron C18R–C19N transgression, although it is not particularly striking on the Vail curves (Haq et al. 1987, 1988). In fact, Chron C19 is well known as an episode of global plate reorganization, when many old spreading ridges died and new spreading ridges (such as the one between Australia and Antarctica) developed (Rona and Richardson 1978).

How could this tectonic change cause cooling? Bartek et al. (1992) argued that Antarctica, which has been situated on the South Pole since early in the Mesozoic, was probably always cold enough for ice caps. Using computer models of climate, they showed that the limiting factor was not temperature, but precipitation. Under the Cretaceous and early Cenozoic conditions of continental climates on the Australia-Antarctica landmass, with Antarctic waters circulating with warmer waters, there was not enough moisture passing over the South Pole to form ice caps. When a seaway opened up between Australia and Antarctica, it moderated the continental climate of the region and allowed onshore transport of moisture over Antarctica (figure 6.5A). There was also a shallow seaway between Marie Byrd Land (on west Antarctica) and East Antarctica (figure 5.2), presumably filling the subglacial Pensacola and Wilkes basins, which separate the two halves of the continent (Harwood 1991). Climatic modeling (Bartek et al. 1992) suggests that air flowing away from the Polar High on East Antarctica over this seaway picked up moisture, which then precipitated as snow when it rose over the mountains of Marie Byrd Land (figure 5.2). Ultimately, early Oligocene ice sheets developed in this region (LeMasurier and Rex 1982; Zachos et al. 1992).

These models are particularly attractive for explaining the Chron C18R–C19N event at the end of the middle Eocene, as this was a time of rapid widening of the seaway between the continents. However, it is more difficult to explain the cooling at the beginning of the middle Eocene with these climatic models, because there were no significant changes in Antarctic seaways that we know of. Other than the isotopic evidence of cooling, the only striking oceanographic change at the end of the early Eocene was a widespread episode of silica deposition. Called the "silica burp" by Brian McGowran (1989), it was apparently the product of deep weathering of silicate minerals during the tropical early Eocene, which led to leaching of silica that eventually washed into the sea. While oceanic waters remained warm, this silica could remain in solution, but the middle Eocene cooling triggered massive chemical deposition of silica all over the world's oceans. The "silica burp" was not, however, a cause of the cooling, but an effect, so the explanation must lie somewhere else.

McGowran (1989; McGowran et al. 1992) has suggested that the beginning of the middle Eocene might reflect the transition between the early Eocene "greenhouse" earth and a world with less atmospheric CO_2. He points to a similarity of isotopic patterns with those that occurred in the middle Miocene, when a warm earth was rapidly cooled and the present-day Antarctic ice sheets developed. According to the "Monterey hypothesis" of Vincent and Berger (1985; Berger and Vincent 1986), an early Miocene enrichment in ^{13}C was followed by heavy carbon depletion; this was followed shortly by an enrichment in ^{18}O, signaling the ice buildup. According to the "Monterey hypothesis," this isotopic pattern signals a "reverse greenhouse"—carbon dioxide leaving the greenhouse gases of the atmosphere and being absorbed by oceanic reservoirs. McGowran (1989) argued that a similar pattern can be seen in the stable isotopic ratios during the Eocene. Throughout the early Eocene and the beginning of the middle Eocene, there is about a 1.5‰ increase in the $\delta^{13}C$, suggesting that the early Eocene "greenhouse" was deteriorating. As noted previously, the beginning of the middle Eocene experienced a 1‰ increase in $\delta^{18}O$, which marked the beginning of the Eocene cooling.

The Greenhouse of the Dinosaurs—Once Again

To understand the cooling and climatic deterioration of the middle Eocene through Oligocene, ultimately we must understand why it became so warm in the first place. The early Eocene was extraordinarily warm and balmy at almost all latitudes. Since the acceptance of plate tectonics, the prevailing explanation for the Cenozoic cooling trend has been the distribution of continents, with the Antarctic land mass settling over the South Pole, thence controlling ice volume and ultimately temperature (Crowell and Frakes 1970; Frakes and Kemp 1972, 1973). Using computer simulations of climate, however, Eric Barron (1985) was able to show that continental position was not enough to explain early Eocene warming. Some other factor, possibly excess greenhouse gases, would be required to account for such warmth. According to some geochemical estimates, atmospheric CO_2 was eight times the present value in the early Eocene, and was still four times modern levels in the late Oligocene (Arthur et al. 1991; Freeman and Hayes 1992).

Where did early Eocene greenhouse gases come from? The most likely source was volcanic gases from the mantle, and several authors (Berner et al. 1983; Owen and Rea 1985) have shown that high rates of sea-floor spreading and ridge volcanism are associated with greenhouse conditions, especially in the Cretaceous. The beginning of the Eocene was marked by massive eruptions of flood basalts in the North Atlantic, producing about 2 million cubic kilometers of basalts (Roberts et al. 1984), only slightly less than the great Deccan eruptions in India that coincided with the end of the Cretaceous. Indeed, the entire North Atlantic began to spread much more rapidly in the early Eocene as new ridges appeared. There was also widespread plate reorganization in the Pacific and Indian Oceans; the collision of India with Asia forced some oceanic ridges to shut down and others to spring up (Williams 1986). In short, early Eocene Chron C24R was marked by a wide variety of plate reorganization events around the world that could have significantly increased mantle eruptions and greenhouse gases (Rona and Richardson 1978; Williams 1986).

Some paleoclimatologists are unhappy with explanations that tender carbon dioxide as the primary greenhouse gas. According to their calculations, a strictly CO_2 greenhouse during the early Eocene would have warmed not only the polar regions, but also the tropics. As we have seen, the tropics were warm, but not much warmer than they are today. Sloan et al.

(1992) have suggested that some of the early Eocene greenhouse might have been caused by methane in addition to CO_2. Methane gas (CH_4, also known as swamp gas, and the major gas burned as "natural gas") is produced in a number of environments, but in the Eocene, the major source would have been swamps and wetlands. Indeed, Paleocene and Eocene coal swamp deposits are well known all over the world. Sloan et al. (1992) argue that rising sea levels in the late Paleocene and early Eocene might have flooded large areas of continental shorelines, and thereby greatly increased the atmospheric methane levels. Methane is not only a good greenhouse gas, but unlike CO_2, it can form optically thick stratospheric clouds in the polar regions, which would have enhanced the warming in the poles without warming the equatorial region.

Another problem with the CO_2 greenhouse comes from recent isotopic studies of sediments in deep-sea cores from the Paleocene-Eocene transition. According to Lowell Stott (1992), the atmospheric CO_2 content appears to have *decreased* slightly for about 10,000 years during the Paleocene-Eocene transition. This idea is so new that no one has had a chance to evaluate his models in the context of other oceanic cores. If Stott's data and interpretations hold up, however, then the conventional ideas about CO_2 greenhouses will have to be rethought. Either there was no increase in greenhouse gases (which most atmospheric scientists would not accept), or the ocean was not absorbing CO_2 in the usual manner. According to Stott (1992), there is no evidence that the ocean took up more CO_2 in the traditional reservoirs of organic carbon or calcium carbonate in limestones in the oceans. Instead, Stott suggests that oceanic circulation might have been unusually sluggish and did not absorb as much CO_2 as conventional estimates dictate. Alternatively, Stott suggests that the carbonate chemistry of the ocean might have changed, particularly in the polar regions where the water was unusually warm, so that the ocean had a reduced capacity to absorb CO_2.

The Paleocene-Eocene transition is marked by some unusually rapid events that seem hard to explain by sudden increases in oceanic spreading alone. According to Kennett and Stott (1991), the transition may have taken less than 6000 years, yet in this relatively short time Antarctic surface waters warmed by almost 8°C to a peak temperature of 10°C, and the equatorial regions reached 20°C. This end of Paleocene thermal maximum did not persist. Within about 10,000 years, the peak had passed and temperatures cooled slightly over the next 150,000 years. However, through the rest of the early Eocene, temperatures were still higher than they had been in the late Paleocene.

Evidence from planktonic microfossils and their oxygen and carbon isotopes suggests that the late Paleocene ocean had vigorous circulation between surface and deep waters, the latter produced by an Antarctic source (similar to our modern ocean). In the early Eocene, circulation became sluggish, so that the ocean became uniformly warm, with no significant cold deep-water circulation. Most deep-water, warm-loving foraminiferans became extinct at the end of the Paleocene, and only warm-water microplankton thrived in the early Eocene. The warm tropical Tethyan belt seems to have been the source of this warm water.

The cause of this rapid warming in the early Eocene is still a puzzle. According to Jim Kennett and Lowell Stott (1990, 1991), the deep waters formed in the Tethys were warm and saline. Using stable isotope records from the Antarctic (Maud Rise), Kennett and Stott (1990) hypothesized an early Paleogene two-layered ocean, "Proteus," consisting of warm, saline deep waters formed at mid-low (Tethyan) latitudes and overlain by cooler, intermediate waters of primarily mid-latitude origin (figure 6.6). The late Paleogene ocean, "Proto-Oceanus," was thought to have been three-layered with cold, dense deep waters formed in the Antarctic (proto-Antarctic bottom water), a warm, saline intermediate layer of low latitude origin, and surficial cool waters formed in both polar regions. By contrast, the Neogene and modern ocean is characterized by a thermohaline circulation driven by deep waters formed at high latitudes.

What could have caused warm deep tropical waters to suddenly spread to high latitudes? One possible explanation might

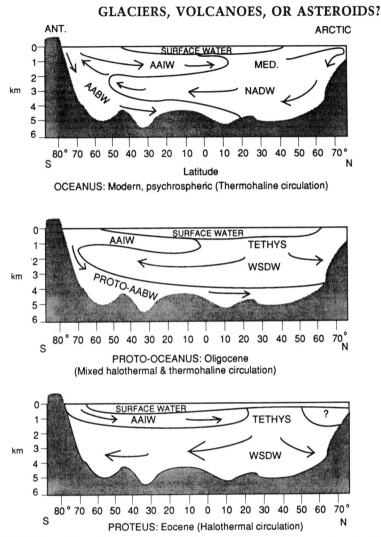

FIGURE 6.6. The "Proteus" and "Proto-Oceanus" hypothesis of oceanic circulation. In the Eocene (bottom), warm saline deep water (WSDW) from tropical Tethys sank due to its salinity and slowly flowed to high latitudes; Antarctic Intermediate Waters (AAIW) flowed north from the southern ocean. In the Oligocene (middle), this pattern began to break up as a proto-Antarctic bottom water (AABW) began to form due to cooling around Antarctica. Since the Oligocene (top), bottom waters come from both the AABW and also the North Atlantic Deep Water (NADW); intermediate waters include the AAIW and the flow out of the Mediterranean (MED). (Kennett and Stott 1990.)

be plate motions. A major tectonic event at the Paleocene-Eocene transition was the collision of India with the Asian continent. The early Eocene also marked the beginning of the closure of the Mediterranean. All of these changes were accompanied by a worldwide acceleration in seafloor spreading rates, which produced much hotter, thicker oceanic crust along the ridge. Since this fresh basalt had not yet cooled and condensed, ridge volume was increased, displacing water out of the ocean basins and causing a rise in sea level. If the tectonic changes along the old Tethyan belt of the Mediterranean and India suddenly enhanced warm Tethyan source waters at the expense of cold southern ocean waters, this might explain the early Eocene warming.

However, some scientists are not convinced. Based on oxygen isotopic records from the Maud Rise, Miller (1992) suggests that at least some of the deep waters in the Southern Ocean were Antarctic in origin, rather than from a tropical warm saline bottom water. Thomas (1992) also argues against the tropics being the sole source of early Eocene bottom waters.

The central paradox about the Paleocene-Eocene warm pulse, and also the early Oligocene cold pulse is their rapidity and short duration. Both took only a few thousand years to develop, and then disappeared after a few hundred thousand years. If these crucial events were controlled by slowly developing processes, such as plate tectonics or atmospheric greenhouse gases, then how do we account for such rapid temperature changes? In addition, such slow processes tend to be unidirectional and irreversible over the short term, so how do we explain the short duration of these thermal events? As discussed above, a number of ideas (rapid opening of oceanic corridors, influence of impacts or volcanic ashes, increases in the seafloor spreading rate and erupted greenhouse gases) have been proposed, but none of these processes operates rapidly enough, or occurs at precisely the right time in the geologic record, to account for the late Paleocene warming or the early Oligocene refrigeration.

Zachos et al. (1993) have suggested that these rapid changes may be examples of threshold effects. Many complex, slowly evolving systems in nature are known to change suddenly when a slight increase in some factor exceeds a critical threshold (Rooth 1982; Crowley and North 1988). Typically, this occurs when a small change in conditions tips the system from one stable equilibrium state into another. For example, it is suggested that the gradual cooling of the Antarctic might not have made any difference until a critical threshold temperature (approximately -10°C) was reached. Below that temperature, Antarctic summers were too cool or too short to completely melt the snow of the previous winter. Once ice sheets persisted, they could expand rapidly if sufficient snow was available. In addition, the beginning of ice-sheet formation has a positive feedback effect, since the ice surface reflects most of the sun's radiation back into the atmosphere. This in turn further cools the polar regions, and increases the snow cover, which increases the reflectivity even further. Such a feedback loop is well known for the Pleistocene, and may also have been critical in the early Oligocene when the first, short-lived ice sheets appeared on Antarctica.

Important data from the Paleocene-Eocene transition is thus still emerging, and there will be much dispute and discussion before any consensus can be agreed upon. However, there is no question that the early Eocene was unusually warm, with sluggish oceanic circulation and probably significant greenhouse gases from volcanoes on rapidly spreading oceanic ridges.

Conclusion

Scientists from many disciplines have provided many pieces of the paleoclimate puzzle, but we do not yet have a complete scenario for the entire history of Eocene atmospheres and oceans. As each new leg of the Oceanic Drilling Project produces new evidence, the story gets more complicated—and more interesting. Nevertheless, we know enough already to draw some important conclusions. Following is the outline of our reconstructed Paleogene climatic history (figure 6.7).

Paleocene-Eocene transition (55 million years ago). The early Eocene was unusually warm, with sluggish oceanic circulation that was warm and saline almost to the bottom. Whether

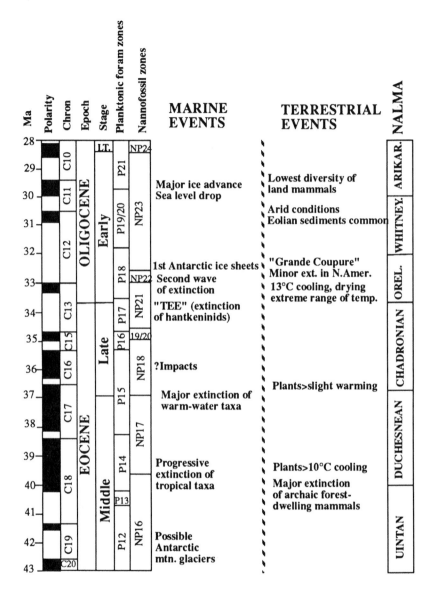

FIGURE 6.7. Summary of marine and terrestrial events in the Eocene and Oligocene.

the source of warm bottom waters was tropical or not, it generated a global climate that was unusually balmy from the equator to high latitudes. A major contributing factor to this warmth may have been greenhouse gases erupting from mid-ocean ridge volcanoes during this episode of rapid spreading, and from massive flood basalt eruptions in the North Atlantic. Whether the collision of India with Asia and the beginning of the closure of the Mediterranean Tethys contributed to these changes is still not clear.

Early-middle Eocene transition (50 million years ago). The beginning of a "reverse greenhouse" seems to have occurred at the end of the early Eocene, with much of the CO_2 in the atmosphere going into carbon reservoirs (mostly coals). The triggers for this are not clear, although most of the rapid spreading of the early Eocene (which had supplied so much greenhouse gas) had ceased. When the carbon began to leave the atmosphere, it produced a "Monterey"-type response, with oxygen isotopes indicating slight cooling at this time. Once cooling had begun, there was "silica burp" in the oceans as dissolved silica precipitated due to cooler waters, as well as mountain glaciers on the Antarctic peninsula.

Middle-late Eocene transition (38-37 million years ago). Further cooling and the biggest extinctions of the Cenozoic seem to coincide with the initial formation of glaciers on Antarctica. Rapid rifting between Antarctica and Australia may have allowed moisture to precipitate as snow over the South Pole.

Mid-late Eocene (35-36 million years ago). Possibly four impacts of extraterrestrial objects occurred, but their effect on climate or extinctions seems to have been negligible. Shallow-water circulation between Antarctica and the South Tasman rise had begun, starting the circum-Antarctic current.

The "Terminal Eocene Event" (34 million years ago). This non-event brought only minor extinctions in planktonic foraminiferans, and relatively little extinction in any other organisms. Most oceanographic events were building up to

The Early Oligocene deterioration (33 million years ago). Here we see solid evidence of massive cooling, some Antarctic glaciation, and deep marine erosion as cold bottom waters flowed into the North Atlantic from the Arctic and enhanced circum-Antarctic circulation stimulated deep water production. This event was responsible for most of the extinctions that were once attributed to the "Terminal Eocene Event."

Mid-Oligocene marine regression (30 million years ago): Major Antarctic glaciation caused the largest sea level drop in the entire Cenozoic, massive deep-sea erosion, and further oceanic cooling. However, the extinction response was not very impressive, possibly because Oligocene organisms were already cold tolerant.

Oligocene-Miocene transition (23 million years ago). A pulse of Antarctic glaciation occurred at the end of the Oligocene, possibly due to the completion of circum-Antarctic circulation through the Drake Passage between South America and Antarctica. However, the global climate began to recover and became warmer during the early Miocene. Diversity of organisms also increased. The modern phase of Antarctic glaciation began in the middle Miocene, and the planet has been in an "icehouse" ever since.

References

Adams, C. G., D. E. Lee, and B. R. Rosen. 1990. Conflicting isotopic and biotic evidence for tropical sea-surface temperatures during the Tertiary. *Palaeogeography, Palaeoclimatology, Palaeoecology* 77: 289–313.

Ager, D. V. 1973. *The Nature of the Stratigraphical Record.* London: Macmillan.

Alvarez, L. W., W. Alvarez, F. Asaro, and H. V. Michel. 1980. Extra-terrestrial cause for the Cretaceous-Tertiary extinction. *Science* 208: 1095–1108.

Alvarez, W., M. A. Arthur, A. G. Fischer, W. Lowrie, G. Napoleone, I. Premoli-Silva and W. R. Roggenthen. 1977. Upper Cretaceous-Paleocene magnetic stratigraphy at Gubbio, Italy: V. Type section for the late Cretaceous–Paleocene geomagnetic reversal time scale. *Geological Society of America Bulletin* 88: 383–389.

Alvarez, W., F. Asaro, H. V. Michel, and L. W. Alvarez. 1982. Iridium anomaly approximately synchronous with terminal Eocene extinctions. *Science* 216: 886–888.

REFERENCES

Andrews, C. W. 1906. *A Descriptive Catalogue of the Tertiary Vertebrata of the Fayum, Egypt.* London: British Museum (Natural History).

Andrews, P., J. M. Lord, and E. M. Nesbit Evans. 1979. Patterns of ecological diversity in fossil and modern mammalian faunas. *Biological Journal of the Linnean Society* 11: 177–205.

Andrews, R. C. 1932. *New Conquest of Central Asia.* New York: American Museum of Natural History.

Arthur, M. A., K. R. Hinga, M. E. Q. Pilson, E. Whitaker, and D. Allard. 1991. Estimates of pCO_2 for the last 120 Ma based on $\delta^{13}C$ of marine phytoplankton organic matter. *EOS, Transactions of the American Geophysical Union* 72: 166.

Asaro, F., L. W. Alvarez, W. Alvarez, and H. V. Michel. 1982. Geochemical anomalies near the Eocene/Oligocene and Permian/ Triassic boundaries. *Geological Society of America Special Paper* 190: 517–528.

Aubry, M. P. 1985. Northwestern European Paleogene magnetostratigraphy, biostratigraphy, and paleogeography: Calcareous nannofossil evidence. *Geology* 13: 198–202.

Aubry, M. P. 1992. Late Paleogene calcareous nannoplankton evolution: A tale of climatic deterioration. In D. R. Prothero and W. A. Berggren, eds., *Eocene-Oligocene Climatic and Biotic Evolution*, pp. 272–309. Princeton: Princeton University Press.

Aubry, M. P., W. A. Berggren, D. V. Kent, J. J. Flynn, K. D. Klitgord, J. D. Obradovich, and D. R. Prothero. 1988. Paleogene geochronology: An integrated approach. *Paleoceanography* 3 (6): 707–742.

Aubry, M. P., F. M. Gradstein, and L. F. Jansa. 1990. The late Early Eocene Montagnais meteorite: No impact on biotic diversity. *Micropaleontology* 36(2): 164–172.

Axelrod, D. I. 1966. The Eocene Copper Basin flora of northeastern Nevada. *University of California Publications in Geological Science* 59: 1–125.

Ba Maw, R. L. Ciochon, and D. E. Savage. 1979. Late Eocene of Burma yields earliest anthropoid primate, *Pondaungia cotteri. Nature* 282: 65–67.

Baldauf, J. G. 1992. Middle Eocene through early Miocene diatom floral turnover. In D. R. Prothero and W. A. Berggren, eds., *Eocene-Oligocene Climatic and Biotic Evolution*, pp. 310–326. Princeton: Princeton University Press.

Baldauf, J. G. and J. A. Barron. 1990. Evolution of biosiliceous sedimentation patterns—Eocene through Quaternary: Paleoceanographic response to polar cooling. In U. Bleil and J. Thiede, eds., *Geological History of the Polar Oceans: Arctic Versus Antarctic*, pp. 575–607. Amsterdam: Kluwer Academic Publishers.

Barker, P. F. and J. Burrell. 1977. The opening of the Drake Passage.

Marine Geology 25: 15–34.

Barker, P. F. and J. Burrell. 1982. The influence upon Southern Ocean circulation sedimentation and climate of the opening of Drake Passage. In C. Craddock, ed., *Antarctic Geoscience*, pp. 377–385. Madison, Wisconsin: University of Wisconsin Press.

Barrett, P. J., M. J. Hambrey, D. M. Harwood, A. R. Pyne, and P. N. Webb. 1989. Synthesis. In P. J. Barrett, ed., *Antarctic Cenozoic History from the CIROS-1 Drillhole, McMurdo Sound (Research Bulletin 245)*, pp. 241–251. Wellington, New Zealand: Science and Information Publishing Centre, Department of Scientific and Industrial Research.

Barron, E. J. 1984. Climatic implications of the variable obliquity explanations of Cretaceous-Paleogene high-latitude floras. *Geology* 12: 595–598.

Barron, E. J. 1985. Explanations of the Tertiary global cooling trend. *Palaeogeography, Palaeoclimatology, Palaeoecology* 50: 45–62.

Barron, J., B. Larsen, and the Leg 119 Shipboard Scientific Party. 1989. *Proceedings of the Ocean Drilling Program* 119: 1–939.

Bartek, L. R. and J. B. Anderson. 1990. Neogene stratigraphy of the Ross Sea continental shelf: Revelations from Leg 2 of the 1990 Ross Sea Expedition of the R/V Polar Duke. In A. K. Cooper and P. N. Webb (conveners), *International Workshop on Antarctic Offshore Seismic Stratigraphy (ANTOSTRAT): Overview and Extended Abstracts. U.S. Geological Survey Open-file Report* 90-309: 53–62.

Bartek, L. R., L. C. Sloan, J. B. Anderson and M. I. Ross. 1992. Evidence from the Antarctic continental margin of late Paleogene ice sheets: A manifestation of plate reorganization and synchronous changes in atmospheric circulation over the emerging Southern Ocean? In D. R. Prothero and W. A. Berggren, eds., *Eocene-Oligocene Climatic and Biotic Evolution*, pp. 131–159. Princeton: Princeton University Press.

Bartels, W. S. 1980. Early Cenozoic reptiles and birds from the Bighorn Basin, Wyoming. *University of Michigan Papers on Paleontology* 24: 73–79.

Berger, W. H. and E. Vincent. 1986. Deep-sea carbonates: Reading the carbon-isotope signal. *Geologische Rundschau* 75: 249–269.

Berggren, W. A. 1971. Tertiary boundaries. In W. R. Riedel and B. M. Funnell, eds., *Marine Micropaleontology of the Oceans*, pp. 693–803. Cambridge: Cambridge University Press.

Berggren, W. A. 1982. Role of ocean gateways in climatic change. In W. Berger and J. C. Crowell, eds., *Climate in Earth History*, pp. 118–285. Washington, D.C.: National Academy of Sciences.

Berggren, W. A. and M. P. Aubry. 1983. Rb-Sr isochron of the Eocene Castle Hayne Limestone, North Carolina—Further dicussion. *Geological Society of America Bulletin* 94: 364–370.

REFERENCES

Berggren, W. A., D. V. Kent and J. J. Flynn. 1985. Paleogene geochronology and chronostratigraphy. In N. J. Snelling, ed., *The Chronology of the Geological Record (Memoir of the Geological Society of London)* 10: 141–195.

Berggren, W. A., D. V. Kent, J. D. Obradovich, and C. C. Swisher III. 1992. Toward a revised Paleogene geochronology. In D. R. Prothero and W. A. Berggren, eds., *Eocene-Oligocene Climatic and Biotic Evolution*, pp. 29–45. Princeton: Princeton University Press.

Berggren, W. A., M. C. McKenna, J. Hardenbol and J. D. Obradovich. 1978. Revised Paleogene polarity time-scale. *Journal of Geology* 86: 67–81.

Berggren, W. A. and D. R. Prothero. 1992. Eocene-Oligocene climatic and biotic evolution: An overview. In D. R. Prothero and W. A. Berggren, eds., *Eocene-Oligocene Climatic and Biotic Evolution*, pp. 1–28. Princeton, NewJersey: Princeton University Press.

Berggren, W. A. and J. A. Van Couvering. 1978. Biochronology. *American Association of Petroleum Geologists Studies in Geology* 6: 39–55.

Berhe, S. M., B. Desta, M. Nicoletti, and M. Teferra. 1987. Geology, geochronology, and geodynamic implications of the Cenozoic magmatic province in W and SE Ethiopia. *Journal of the Geological Society of London* 144: 213–226.

Berner, R. A., A. C. Lasaga, and R. M. Garrels. 1983. The carbonate-silicate geochemical cycle and its effects on atmospheric carbon dioxide over the last 100 million years. *American Journal of Science* 283: 641–683.

Berry, W. B. N. 1987. *Growth of a Prehistoric Time Scale Based on Organic Evolution*, 2nd ed. Palo Alto, California: Blackwell Scientific Publications.

Betts, C. 1871. The Yale College expedition of 1870. *Harper's New Monthly Magazine*, October 1871: 663–671.

Beyrich, H. E. von. 1954. Über die Stellung die hessischen Tertiarbildungen. *K. Preuss. Akad. Wiss. Berlin Monatsheft* 1854: 664–666.

Birkenmajer, K. 1987. Tertiary glacial and interglacial deposits, South Shetland Islands, Antarctica: Geochronology versus biostratigraphy (a progress report). *Bulletin of the Polish Academy Science, Earth Science* 36: 133–145.

Birkenmajer, K. and E. Zastawniak. 1989. Late Cretaceous-Tertiary floras of King Georges Island, West Antarctica: Their stratigraphic distribution and paleoclimatic significance. In J. A. Crame, ed., *Origin and Evolution of the Antarctic Biota. Geological Society of London Special Publication* 47: 227–240.

Black, C. C. and M. R. Dawson. 1966. A review of late Eocene mammalian faunas from North America. *American Journal of Science* 264: 321–349

Boersma, A., I. Premoli-Silva and N. J. Shackleton. 1987. Atlantic Eocene planktonic foraminiferal paleohydrographic indicators and stable isotope paleoceanography. *Paleoceanography* 2(3): 287–331.

Bottomley, R. and D. York. 1988. Age measurements of the submarine Montagnais impact crater and the periodicity question. *Geophysical Research Letters* 14(12): 1409–1412.

Boulter, M. C. 1984. Palaeobotanical evidence for land-surface temperature in the European Paleogene. In P. J. Brenchley, ed., *Fossils and Climate*, pp. 35–47. Chichester: John Wiley and Sons.

Boulter, M. C. and R. N. L. B. Hubbard. 1982. Objective paleo-ecological and biostratigraphic interpretation of Tertiary palyno-logical data by multivariate statistical analysis. *Palynology* 6: 55–68.

Bown, T. M. and M. J. Kraus. 1981. Lower Eocene alluvial paleosols (Willwood Formation, northwest Wyoming, U.S.A.) and their significance for paleoecology, paleoclimatology, and basin analysis. *Palaeogeography, Palaeoclimatology, Palaeoecology* 34: 1–30.

Bown, T. M. and M. J. Kraus. 1987. Integration of channel and floodplain suites: 1, Developmental sequence and lateral relations of alluvial paleosols. *Journal of Sedimentary Petrology* 57: 587–601.

Breza, J., S. W. Wise, J. C. Zachos, and The Leg 120 Shipboard Party. 1989. Lower Oligocene ice-rafted debris at 58°S on the Kerguelen Plateau: The 'Smoking Gun' for the existence of an early Oligocene ice sheet on East Antarctica: *Third International Conference on Paleoceanography, Abstract Volume 4*, p. 24. Cambridge: Blackwell Scientific Publications.

Brinkhuis, H. 1992. Late Paleogene dinoflagellate cysts with special reference to the Eocene/Oligocene boundary. In D. R. Prothero and W. A. Berggren, eds., *Eocene-Oligocene Climatic and Biotic Evolution*, pp. 327–340. Princeton: Princeton University Press.

Brown, R. W. 1950. An Oligocene evergreen cherry from Oregon. *Journal of the Washington Academy of Sciences.* 40: 321–324.

Brunet, M. 1977. Les mammifères et le probleme de la limite Eocène-Oligocène en Europe. *Mémoires Spéciaux Geobios* 1: 11–27.

Buchardt, B. 1978. Oxygen isotope paleotemperatures from the Tertiary period in the North Sea area. *Nature* 275: 121–123.

Bukry, D. and Snavely, P. D. 1988. Coccolith zonation for Paleogene strata in the Oregon Coast range. In Filewicz, M. V. and R. L. Squires, eds., *Paleogene stratigraphy, West Coast of North Ameri ca, Pacific Section, S.E.P.M. West Coast Paleogene Symposium* 58: 251–263.

REFERENCES

Byerly, G. R., J. E. Hazel and C. McCabe. 1990. Discrediting the late Eocene microspherule layer at Cynthia, Mississippi. *Meteoritics* 25: 89–92.

Callahan, J. E. 1971. Velocity structure and flux of the Antarctic Circumpolar Current of South Australia. *Journal of Geophysical Research* 76: 5859–5870.

Cande, S. C. and D. V. Kent. 1992. A new geomagnetic polarity timescale for the late Cretaceous and Cenozoic. *Journal of Geophysical Research* 97 (B10): 13917-13951,

Capetta, H., J. J. Jaeger, M. Sabatier, B. Sigé, J. Sudre and M. Vianey-Liaud. 1978. Découverte dans le Paléocène du Maroc des plus anciens mammifères euthériens d'Afrique. *Géobios* 11: 257–262.

Case, J. A. 1988. Paleogene floras from Seymour Island, Antarctic Peninsula. *Geological Society of America Memoir* 169: 523–530.

Cavelier, C. 1979. La limite Eocène-Oligocène en Europe occidentale. *Sci. Géol. Inst. Géol. Strasbourg, (Mém.)* 54: 1–280.

Chaney, R. W. and E. I. Sanborn. 1933. The Goshen flora of west central Oregon. *Carnegie Institute of Washington Publication* 439: 1–103.

Chow Minchen and Wang Banyue. 1979. Relationship between the pantodonts and tillodonts and classification of the order Pantodonta. *Vertebrata PalAsiatica* 17: 37–48.

Christophel, D. C. 1990. The impact of mid-Tertiary climatic changes on the development of the modern Australian flora. *Geological Society of America, Abstracts with Programs* 22(7): A77.

Cifelli, R. 1969. Radiation of Cenozoic planktonic Foraminifera. *Systematic Zoology* 18: 154–168.

Cifelli, R. L. 1985. South American ungulate evolution and extinction. In Stehli, F.G. and S. D. Webb, eds., *The Great American Biotic Interchange*, pp. 249–266. New York: Plenum Press.

Ciochon, R., D. Savage, R. Tint, and Ba Maw. 1985. Anthropoid origins in Asia? New discoveries of *Amphipithecus* from the Eocene of Burma. *Science* 229: 756-759.

Cloetingh, S., H. McQueen, and K. Lambeck. 1985. On a tectonic mechanism for regional sea level variations. *Earth and Planetary Science Letters* 75: 157–166.

Coffin, H. 1976. Orientations of trees in the Yellowstone petrified forests. *Journal of Paleontology* 50: 539–543.

Colbert, E. H. 1938. Fossil mammals from Burma in the American Museum of Natural History. *Bulletin of the American Museum of Natural History* 74: 255–436.

Collinson, M. E. 1983. Palaeofloristic assemblages and palaeoecology of the lower Oligocene Bembridge Marls, Hamstead Ledge, Isle of Wight. *Botanical Journal of the Linnean*

254

Society 86: 177–225.

Collinson, M. E. 1992. Vegetational and floristic changes around the Eocene/Oligocene boundary in western and central Europe. In D. R. Prothero and W. A. Berggren, eds., *Eocene-Oligocene Climatic and Biotic Evolution*, pp. 437–450. Princeton: Princeton University Press.

Collinson, M. E., K. Fowler and M. C. Boulter. 1981. Floristic changes indicate a cooling climate in the Eocene of southern England. *Nature* 291: 315–317.

Collinson, M. E. and J. J. Hooker. 1987. Vegetational and mammalian faunal changes in the early Tertiary of southern England. In E. M. Friis, W. G. Chaloner and P. R. Crane, eds., *The Origins of Angiosperms and their Biological Consequences*, pp. 259–304. Cambridge: Cambridge University Press.

Corliss, B. H., M. P. Aubry, W. A. Berggren, J. M. Fenner, L. D. Keigwin, and G. Keller. 1984. The Eocene/Oligocene boundary event in the deep sea. *Science* 226: 806–810.

Corliss, B. H. and L. D. Keigwin. 1986. Eocene-Oligocene paleoceano-graphy. *American Geophysical Union Geodynamics Series* 15: 101–118.

Courtillot, V. E. 1990. Volcanic eruptions. *Scientific American* 263(4): 85–92.

Cox, A. 1969. Geomagnetic reversals. *Science* 163: 237–244.

Creber, G. T. and W. G. Chaloner. 1984. Climatic indications from growth rings in fossil woods. In P. J. Brenchley, ed., *Fossils and Climate*, pp. 49–74. New York: John Wiley and Sons.

Crowell, J. and L. A. Frakes. 1970. Phanerozoic glaciation and the causes of the ice ages. *American Journal of Science* 268: 193–224.

Crowley, T. J. and G. R. North. 1988. Abrupt climate change and extinction events in earth history. *Science* 240: 996–1002.

Curry, D. and G. S. Odin. 1982. Dating of the Paleogene. In G. S. Odin, ed., *Numerical Dating in Stratigraphy*, pp. 607–630. New York: John Wiley.

D'Hondt, S. L., G. Keller, and R. F. Stallard. 1987. Major element compositional variation within and between different late Eocene microtektite strewnfields. *Meteoritics* 22: 61–79.

Dashvezeg, D. and E. V. Devyatkin. 1986. Eocene–Oligocene boundary in Mongolia. In C. Pomerol and I. Premoli-Silva, eds., *Terminal Eocene Events*, pp. 153–157. Amsterdam: Elsevier.

Dashzeveg, D. and D. E. Russell. 1988. Paleocene and Eocene Mixodontia (Mammalia, Glires) of Mongolia and China. *Palaeontology* 31: 129–164.

Davies, A. M. 1934. *Tertiary Faunas*, 2 vols. London: Thomas Murby & Co.

Davis, M., P. Hut, and R. A. Muller. 1984. Extinction of species by periodic comet showers. *Nature* 308: 715–717.

REFERENCES

Devereux, I. 1967. Oxygen isotope paleotemperature measurements on New Zealand Tertiary fossils. *New Zealand Journal of Science* 10: 988–1011.

Donnelly, T. W. and E. C. T. Chao. 1973. Microtektites of late Eocene age from the eastern Caribbean Sea. *Initial Reports of the Deep Sea Drilling Project* 7: 607–672.

Dorf, E. 1960. Tertiary fossil forests of Yellowstone National Park, Wyoming. *Billings Geological Society Annual Field Conference* 11: 253–360.

Dorf, E. 1964. The petrified forests of Yellowstone National Park. *Scientific American* 210(4): 106–114.

Dumont, A. 1839. Rapport sur les travaux de la carte géologique en 1839, avec une carte géologique des environs de Bruxelles. *Bulletin de Academie Royale de Belgique* 6(2): 464–485.

Dumont, A. 1849. Rapport sur la carte géologique du Royaume. *Bulletin de Academie Royale de Belgique.* 16(2): 351–373.

Emiliani, C. 1954. Depth habitats of some species of pelagic foraminifera as indicated by oxygen isotope ratios. *American Journal of Science* 252: 149–158.

Emry, R. J. 1973. Stratigraphy and preliminary biostratigraphy of the Flagstaff Rim area, Natrona County, Wyoming. *Smithsonian Contributions to Paleobiology* 18.

Emry, R. J. 1981. Additions to the mammalian fauna of the type Duchesnean, with comments on the status of the Duchesnean. *Journal of Paleontology* 55: 563–570.

Emry, R. J., P. R. Bjork, and L. S. Russell. 1987. The Chadronian, Orellan, and Whitneyan land mammal ages. In M. O. Woodburne, ed., *Cenozoic Mammals of North America, Geochronology and Biostratigraphy*, pp. 118–152. Berkeley: University of California Press.

Erwin, D. J. 1993. *The Great Paleozoic Crisis: Life and Death in the Permian.* New York: Columbia University Press.

Estes, R. and J. H. Hutchinson. 1980. Eocene lower vertebrates from Ellesmere Island, Canadian Arctic Archipelago. *Palaeogeography, Palaeoclimatology, Palaeoecology* 30: 325–347.

Evanoff, E., D. R. Prothero and R. H. Lander. 1992. Eocene-Oligocene climatic change in North America: The White River Formation near Douglas, east-central Wyoming. In D. R. Prothero and W. A. Berggren, eds., *Eocene-Oligocene Climatic and Biotic Evolution*, pp. 116–130. Princeton, NewJersey: Princeton University Press.

Evernden, J. F., D. E. Savage, G. H. Curtis and G. T. James. 1964. Potassium-argon dates and the Cenozoic mammalian chronology of North America. *American Journal of Science* 262: 145–198.

Fenner, J. 1986. Information from diatom analysis concerning the Eocene-Oligocene boundary. In C. H. Pomerol and I. Premoli-Silva, eds., *Terminal Eocene Events*, pp. 283-288. Amsterdam:

Elsevier.

Filhol, H. 1876. Recherches sur les phosphorites du Quercy. Etude des fossiles qu'on y rencontre et spécialement des Mammifères. *Annales Sciences Géologiques* 7(7): 1–220.

Flynn, J. J. 1986. Correlation and geochronology of middle Eocene strata from the western United States. *Palaeogeography, Palaeoclimatology, Palaeoecology* 55: 335–406.

Flynn, J. J. and A. R. Wyss. 1990. New early Oligocene marsupials from the Andean Cordillera, Chile. *Journal of Vertebrate Paleontology* 10(3): 22A (abstract).

Fordyce, R. E. 1980. Whale evolution and Oligocene Southern Ocean environments. *Palaeogeography, Palaeoclimatology, Palaeoecolo-gy* 31: 319–336.

Fordyce, R. E. 1989. Origins and evolution of Antarctic marine mammals. *Special Publications of the Geological Society of London* 47: 269–281.

Fordyce, R. E. 1992. Cetacean evolution and Eocene-Oligocene environments. In D. R. Prothero and W. A. Berggren, eds., *Eocene-Oligocene Climatic and Biotic Evolution*, pp. 368–381. Princeton: Princeton University Press.

Frakes, L. A. 1979. *Climates Throughout Geological Time*. New York: Elsevier.

Frakes, L. A. and E. M. Kemp. 1972. Influence of continental positions on early Tertiary climates. *Nature* 240: 97–100.

Frakes, L. A. and E. M. Kemp. 1973. Palaeogene continental positions and evolution of climate. In D. H. Tarling and S. K. Runcorn, eds., *Implications of Continental Drift to the Earth Sciences* 1: 541–559. New York: Academic Press.

Franzen, J. L. 1987. Ein neuer Primate aus dem Mitteleozän der Grube Messel (Deutschland, Sud-Hessen). *Courier Forschunginstitut Senckenberg* 91: 151–187.

Frederiksen, N. O. 1988. *Sporomorph Biostratigraphy, Floral Changes, and Paleoclimatology, Eocene and Earliest Oligocene of the Eastern Gulf Coast, Professional Paper 1448*. United States Geological Survey.

Frederiksen, N. O. 1991. Pulses of middle Eocene to earliest Oligocene climatic deterioration in southern California and the Gulf Coast. *Palaios* 6: 564–571.

Freeman, K. H. and J. M. Hayes. 1992. Fractionation of carbon isotopes by phytoplankton and estimates of ancient CO_2 levels. *Global Biogeochemical Cycles* 6: 185–198.

Fritz, W. J. 1980. Reinterpretation of the depositional environments of the Yellowstone "fossil forests." *Geology* 8: 309–313.

Fritz, W. J. 1986. Plant taphonomy in areas of explosive vulcanism. In T. W. Broadhead, ed., *Land Plants: Notes for a short course. University of Tennessee Department of Geological Sciences*

REFERENCES

Studies in Geology 15: 1–9.

Ganapathy, R. 1982. Evidence for a major meteorite impact on the Earth 34 million years ago: Implication for Eocene extinctions. *Science* 216: 885–886.

Gaskell, B. A. 1991. Extinction patterns in Paleogene benthic foraminiferal faunas: Relationship to climate and sea level. *Palaios* 6: 2–16.

Gazin, C. L. 1968. A study of the Eocene condylarthran mammal *Hyopsodus*. *Smithsonian Miscellaneous Collections* 149 (2): 1-98.

Gheerbrant, E. 1987. Les vertébrés continentaux de l'Adrar Mgorn (Maroc, Paléocène); une dispersion de mammifères transtéthysienne aux environs de la limite mésozoique/cénozoique? *Geodinamica Acta* 1: 233–246.

Gingerich, P. D. 1976. Cranial anatomy and evolution of early Tertiary Plesiadapidae (Mammalia, Primates). *University of Michigan Papers on Paleontology* 15: 1–141.

Gingerich, P. D. 1977. Radiation of Eocene Adapidae in Europe. *Géobios, Mémoire Spécial* 1: 165–182.

Gingerich, P. D. 1985. South American mammals in the Paleocene of North America. In F. G. Stehli and S. D. Webb, eds., *The Great American Biotic Interchange*, pp. 123–137. New York: Plenum Press.

Gingerich, P. D. 1986. Early Cenozoic *Cantius torresi*:: Oldest primate of modern aspect from North America. *Nature* 319: 319–321.

Gingerich, P. D. 1989. New earliest Wasatchian mammalian fauna from the Eocene of northwestern Wyoming: Composition and diversity in a rarely sampled high-floodplain assemblage. *University of Michigan Papers on Paleontology* 28: 1–97.

Gingerich, P. D. and G. F. Gunell. 1979. Systematics and evolution of the genus *Esthonyx* (Mammalia, Tillodontia) in the early Eocene of North America. *University of Michigan Papers on Paleontology* 25: 125–153.

Gingerich, P. D., B. H. Smith, and E. L. Simons. 1990. Hind limbs of Eocene *Basilosaurus*: Evidence of feet in whales. *Science* 249: 154–157.

Gingerich, P. D., N. A. Wells, D. E. Russell, and S. M. Ibrahim Shah. 1983. Origin of whales in epicontinental remnant seas; new evidence from the early Eocene of Pakistan. *Science* 220: 403–406.

Glass, B. P. 1974. High-magnesium glasses associated with North American microtektites in a Caribbean deep-sea sediment core. *Meteoritics* 9: 345–347.

Glass, B. P. 1984. Multiple microtektite horizons in upper Eocene marine sediments: Comment. *Science* 224: 309.

Glass, B. P. 1986. Late Eocene microtektites and clinopyroxene-bearing spherules. In C. Pomerol and I. Premoli-Silva, eds., *Terminal Eocene Events*, pp. 395–401. Amsterdam: Elsevier.

Glass, B. P. 1990. Chronostratigraphy of upper Eocene microspherules: Comment. *Palaios* 5: 387–389.

Glass, B. P., R. N. Baker, R. N. Storzer, and G. A. Wagner. 1973. North American microtektites from the Caribbean Sea and Gulf of Mexico. *Earth and Planetary Science Letters* 19: 184–192.

Glass, B. P., C. A. Burns, J. R. Crosbie, and D. L. DuBois. 1985. Late Eocene North American microtektites and clinopyroxene-bearing spherules. *Journal of Geophysical Research* 90: D175–D196.

Glass, B. P. and J. R. Crosbie. 1982. Age of the Eocene/Oligocene boundary based on extrapolation from North American microtektite layer. *Bulletin of the American Association of Petroleum Geologists* 66: 471–476.

Glass, B. P., D. L. DuBois, and R. Ganapathy. 1982. Relationship between an iridium anomaly and the North American micro-tektite layer in core RC9-58 from the Caribbean Sea. *Journal of Geophysical Research* 87: 425–428.

Glass, B. P., C. D. Hall, and D. York. 1986. $^{40}Ar/^{39}Ar$ laser-probe dating of North American tektite fragments from Barbados and the age of the Eocene-Oligocene boundary. *Chemical Geology (Isotope Geoscience Section)* 59: 181–186.

Glass, B. P., M. B. Swincki, and P. A. Zwart. 1979. Australasian, Ivory Coast, and North American tektite strewn fields: size, mass, and correlation with geomagnetic reversals and other earth events. *Proceedings of the Lunar and Planetary Science Conference* 10: 2535–2545.

Glass, B. P. and M. J. Zwart. 1977. North American microtektites, radiolarian extinctions and the age of the Eocene/Oligocene boundary. In F. M. Swain, ed., *Stratigraphic Micropaleontology of the Atlantic Basin and Borderlands*, pp. 553–568. Amsterdam: Elsevier.

Glass, B. P. and M. J. Zwart. 1979. North American micro-tektites in the Deep Sea Drilling Project cores from the Caribbean Sea and Gulf of Mexico. *Geological Society of America Bulletin* 90: 595–602.

Godthelp, H., M. Archer, R. L. Cifelli, S. J. Hand, and C. F. Gilkerson. 1992. Earliest known Australian Tertiary mammal fauna. *Nature* 356: 514–516.

Goodney, D. E., S. V. Margolis, W. C. Dudley, P. Kroopnick, and D. F. Williams. 1980. Oxygen and carbon isotopes of Recent calcareous nannofossils as paleoceanographic indicators. *Marine Micropaleontology* 5: 31–42.

Grande, L. 1980. Paleontology of the Green River Formation, with a review of the fish fauna. *Geological Survey of Wyoming Bulletin* 63: 1–333.

REFERENCES

Granger, W. and W. K. Gregory. 1943. A revision of the Mongolian titanotheres. *Bulletin of the American Museum of Natural History* 80: 349–389.

Gregory, J. T. 1971. Speculations of the significance of fossil vertebrates for the antiquity of the High Plains in North America. *Abh. Hessisches Landesamtes für Bodenforsch. (Heinz Tobien Festscrift)*, pp. 64–72.

Hallam, A. 1984. The causes of mass extinctions. *Nature* 308: 686–687.

Hallam, A. 1988. A reevaluation of Jurassic eustasy in light of new data and the revised Exxon curve. *S.E.P.M. Special Publication* 42: 261–274.

Hallam, A. 1992. *Phanerozoic Sea Level Changes*. New York: Columbia University Press.

Hansen, T. A. 1987. Extinction of late Eocene to Oligocene molluscs: Relationship to shelf area, temperature changes, and impact events. *Palaios* 2: 69–75.

Hansen, T. A. 1988. Early Tertiary radiation of marine molluscs and the long-term effects of the Cretaceous-Tertiary extinction. *Paleobiology* 14: 37–51.

Hansen, T. A. 1992. The patterns and causes of molluscan extinction across the Eocene/Oligocene boundary. In D. R. Prothero and W. A. Berggren, eds., *Eocene-Oligocene Climatic and Biotic Evolution*, pp. 341–348. Princeton: Princeton University Press.

Hanson, C. B. 1989. *Teletaceras radinskyi*, a new primitive rhinocerotid from the late Eocene Clarno Formation of Oregon. In D. R. Prothero and R. M. Schoch, eds., *The Evolution of Perissodactyls*, pp. 379–398. New York: Oxford University Press.

Haq, B. U., J. Hardenbol, and P. R. Vail. 1987. The chronology of fluctuating sea level since the Triassic. *Science* 235: 1156–1167.

Haq, B. U., J. Hardenbol, and P. R. Vail. 1988. Mesozoic and Cenozoic chronostratigraphy and cycles of sea-level change. In C. K. Wilgus, H. R. Posamentier, C. A. Ross and C. G. Kendall,, eds., *Sea Level Changes: An Integrated Approach. S.E.P.M. Special Publication* 42: 71–108.

Haq, B. U. and G. P. Lohmann. 1976. Early Cenozoic calcareous nannoplankton biogeography of the Atlantic Ocean. *Marine Micropaleontology* 1: 119–194.

Haq, B. U., I. Premoli-Silva, and G. P. Lohmann. 1977. Calcareous plankton paleobiogeographic evidence for major climatic fluctuations in the early Cenozoic Atlantic Ocean. *Journal of Geophysical Research* 82: 3861–3876.

Hardenbol, J. and W. A. Berggren. 1978. A new Paleogene numerical time scale. *American Association of Petroleum Geologists Studies in Geology* 6: 213–234.

Harper, C. W., Jr. 1987. Might Occam's canon explode the Death

Star?: A moving average model of biotic extinctions. *Palaios* 2: 600–604.

Harris, A. W. and W. R. Ward. 1982. Dynamical constraints on the formation and evolution of planetary bodies. *Annual Review of Earth and Planetary Sciences* 10: 61–108.

Harris, W. B. and P. D. Fullagar. 1989. Comparison of Rb-Sr and K-Ar dates of middle Eocene bentonite and glauconite, southeastern Atlantic Coastal Plain. *Geological Society of America Bulletin* 101(4): 573–577.

Harris, W. B., P. D. Fullagar, and J. A. Winters. 1984. Rb-Sr glauconite ages, Sabinian, Claibornian, and Jacksonian units, southeastern Atlantic Coastal Plain, U.S.A. *Palaeogeography, Palaeoclimatology, Palaeoecology* 47: 53–76.

Harris, W. B. and V. A. Zullo. 1980. Rb-Sr glauconite isochron of the Eocene Castle Hayne Limestone, North Carolina. *Geological Society of America Bulletin* 93: 587–592.

Harrison, C. G. A., I. McDougall, and N. D. Watkins. 1979. A geomagnetic field reversal time scale back to 13.0 million years before present. *Earth and Planetary Sciences Letters* 42: 143–152.

Hartenberger, J. L. 1986. Crises biologiques en milieu continental au cours du Paléogène: exemple des mammifères d'Europe. *Bull. Centr. Rech., Expl. Prod. Elf-Aquitaine* 10: 489–500.

Hartenberger, J. L. 1987. Modalités des extinctions et apparitions chez les mammifères du Paléogène d'Europe. *Memoir de Societe Géologie de France, N.S.* 150: 133–143.

Hartenberger, J. L. 1988. Etudes sur la longevité des genres mammifères fossils du Paléogène d'Europe. *Comptes Rendus Hebdomadaires des Seances de l'Academie des Sciences* 306: 1197–1204.

Hartenberger, J. L., C. Martinez, and A. Ben Said. 1985. Decouverte de mammifères d'age Éocène inférieur en Tunisie Centrale. *Comptes Rendus Hebdomadaires des Seances de l'Academie des Sciences* 301: 649–652.

Hazel, J. E. 1989. Chronostratigraphy of upper Eocene microspherules. *Palaios* 4: 318–329.

Hazel, J. E. 1990. Chronostratigraphy of upper Eocene microspherules: Reply. *Palaios* 4: 389–390.

Heirtzler, J. R., G. O. Dickson, E. M. Herron, W. C. Pitman III, and X. Le Pichon. 1968. Marine magnetic anomalies, geomagnetic field reversals, and motions of the ocean floor and continents. *Journal of Geophysical Research* 73: 2119–2136.

Heissig, K. 1979. Die hypothetische Rolle Sudosteuropas bei den Säugetierwanderungen im Eozän und Oligozän. *Neues Jahrbuch Geologische und Paläontologische Monatsheft* 1979: 83–96.

Heissig, K. 1986. No effect of the Ries impact event on the local mammal fauna. *Modern Geology* 10: 171–179.

261

REFERENCES

Hickey, L. J. 1977. Stratigraphy and paleobotany of the Golden Valley Formation (early Tertiary) of western North Dakota. *Geological Society of America Memoir* 150: 1–181.

Hillhouse, J. W. and C. S. Grommé. 1982. Limits to the northwest drift of the Paleocene Cantwell Formation, central Alaska. *Geology* 14: 552–556.

Hoffman, A. 1985. Patterns of family extinction: dependence on definition and geologic time scale. *Nature* 315: 659–662.

Hoffman, A. 1989a. *Arguments on Evolution: A Paleontologist's Perspective.* New York: Oxford University Press.

Hoffman, A. 1989b. Mass extinction: the view of a sceptic. *Journal of the Geological Society of London* 146: 21–35.

Hoffman, A. and J. Ghiold. 1986. Randomness in the pattern of 'mass extinctions' and 'waves of originations.' *Geological Maga-zine* 122: 1–4.

Hooker, J. J. 1989. Character polarities in early perissodactyls and their significance for *Hyracotherium* and infraordinal relationships. In D. R. Prothero and R. M. Schoch, eds., *The Evolution of Perissodactyls*, pp. 79–101. New York: Oxford University Press.

Hooker, J. J. 1992. British mammalian paleocommunities across the Eocene-Oligocene transition and their environmental implications. In D. R. Prothero and W. A. Berggren, eds., *Eocene-Oligocene Climatic and Biotic Evolution*, pp. 494–515. Princeton: Princeton University Press.

Hubbard, R. J. 1988. Age and significance of sequence boundaries on Jurassic and Early Cretaceous rifted continental margins. *American Association of Petroleum Geologists Bulletin* 72: 49–72.

Hubbard, R. N. L. B. and M. C. Boulter. 1983. Reconstruction of Palaeogene climate from palynological evidence. *Nature* 301: 147–150.

Hull, D. 1988. *Science as a Process.* Chicago: University of Chicago Press.

Hut, P., W. Alvarez, W. P. Elder, T. Hansen, E. G. Kauffman, G. Keller, E. M. Shoemaker, and P. Weismann. 1987. Comet showers as a cause of mass extinctions. *Nature* 329: 118–126.

Hutchison, J. H. 1982. Turtle, crocodilian and champsosaur diversity changes in the Cenozoic of the north-central region of the western United States. *Palaeogeography, Palaeoclimatology, Palaeoecology* 37: 149–164.

Hutchison, J. H. 1992. Western North American reptile and amphibian record across the Eocene/Oligocene boundary and its climatic implications. In D. R. Prothero and W. A. Berggren, eds., *Eocene-Oligocene Climatic and Biotic Evolution*, pp. 451–463. Princeton: Princeton University Press.

Jansa, L. F., M. P. Aubry, and F. M. Gradstein. 1990. Comets and extinctions: Cause and effect? *Geological Society of America Special Paper* 247: 223–232.

Jouse, A. P. 1978. Diatom biostratigraphy on the generic level. *Micropaleontology* 24: 316–326.

Kaminski, M. A. 1987. Cenozoic deep-water agglutinated foraminifera in the North Atlantic. PhD Thesis, Massachusetts Institute of Technology/Woods Hole Oceanographic Institute 88-3, 262 pp.

Kamp, P. J. J., D. B. Waghorn, and C. S. Nelson. 1990. Late Eocene–early Oligocene integrated isotope stratigraphy and biostratigraphy for paleoshelf sequences in southern Australia: Paleoceanographic implications. *Palaeogeography, Palaeoclimatology, Palaeoecology* 80: 311–323.

Keigwin, L. D. and B. H. Corliss. 1986. Stable isotopes in late middle Eocene to Oligocene foraminifera. *Geological Society of America Bulletin* 97: 335–345.

Keigwin, L. D. and G. Keller. 1984. Middle Oligocene climate change from equatorial Pacific DSDP Site 77. *Geology* 12: 16–19.

Keller, G. 1983a. Paleoclimatic analyses of middle Eocene through Oligocene planktic foraminiferal faunas. *Palaeogeography, Palaeo-climatology, Palaeoecology* 43: 73–94.

Keller, G. 1983b. Biochronology and paleoclimatic implications of middle Eocene to Oligocene planktic foraminiferal faunas. *Marine Micropaleontology* 7: 463–486.

Keller, G. 1985. Eocene and Oligocene stratigraphy and erosional unconformities in the Gulf of Mexico and the Gulf Coast. *Journal of Paleontology* 59: 882–903.

Keller, G. 1986a. Late Eocene impact events and stepwise mass extinctions. In C. Pomerol and I. Premoli-Silva, eds., *Terminal Eocene Events*, pp. 403–412. Amsterdam: Elsevier.

Keller, G. 1986b. Stepwise mass extinctions and impact events. *Marine Micropaleontology* 10: 267–293.

Keller, G., S. L. D'Hondt, C. J. Orth, J. S. Gilmore, P. Q. Oliver, E. M. Shoemaker, and E. Molina. 1987. Late Eocene impact microspherules: Stratigraphy, age, and geochemistry. *Meteoritics* 22: 25–60.

Keller, G., S. D'Hondt, and T. L. Vallier. 1983. Multiple micro tektites horizon in upper Eocene marine sediments: No evidence for mass extinctions. *Science* 221: 150–152.

Keller, G., S. D'Hondt, and T. L. Vallier. 1984. Multiple microtektite horizons in upper Eocene marine sediments: Reply. *Science* 224: 309–310.

Keller, G., T. Herbert, R. Dorsey, S. D'Hondt, M. Johnsson, and W. R. Chi. 1987. Global distribution of late Paleogene hiatuses. *Geology* 15: 199–203.

Kelly, T. S. 1990. Biostratigraphy of Uintan and Duchesnean land mammal assemblages from the middle member of the Sespe

Formation, Simi Valley, California. *Contributions to Science of the Natural History Museum of Los Angeles County* 419: 1–42.

Kemp, E. M. 1975. Palynology of Leg 28 drill sites, Deep Sea Drilling Project. *Initial Reports of the Deep Sea Drilling Project* 28: 599–623.

Kemp, E. M. 1978. Tertiary climatic evolution and vegetation history in the southeast Indian Ocean region. *Palaeogeography, Palaeoclimatology, Palaeoecology* 24: 169–208.

Kemp, E. M. and P. J. Barrett. 1975. Antarctic glaciations and early Tertiary vegetation. *Nature* 258: 507–508.

Kennett, J. P. 1977. Cenozoic evolution of Antarctic glaciation, the Circum-Antarctic Ocean, and their impact on global paleoceanography. *Journal of Geophysical Research* 82: 3843–3860.

Kennett, J. P. 1978. The development of planktonic biogeography in the Southern Ocean during the Cenozoic. *Marine Micropaleontology* 3: 301–345.

Kennett, J. P. 1980. Paleoceanographic and biogeographic evolution of the Southern Ocean during the Cenozoic, and Cenozoic microfossil datums. *Palaeogeography, Palaeoclimatology, Palaeoecology* 31: 123–152.

Kennett, J. P. 1983. Paleoceanography: Global ocean evolution. *Reviews of Geophysics and Space Physics* 21: 1258–1274.

Kennett, J. P. and P. F. Barker. 1990. Latest Cretaceous to Cenozoic climate and oceanographic developments in the Weddell Sea, Antarctica: An ocean drilling perspective. *Proceedings of the Ocean Drilling Program* 113(Part B): 937–960.

Kennett, J. P., R. E. Houtz, P. B. Andrews, A. R. Edwards, V. A. Gostin, M. Hahos, M. A. Hampton, D. G. Jenkins, S. V. Margolis, A. T. Ovenshine and K. Perch-Nielsen. 1975. Cenozoic paleoceanography in the southwest Pacific Ocean, Antarctic glaciation and the development of the circum-Antarctic current: *Initial Reports of the Deep Sea Drilling Project* 29: 1155–1169.

Kennett, J. P. and L. D. Stott. 1990. Proteus and Proto-Oceanus: Ancestral Paleogene oceans as revealed from Antarctic stable isotopic results; ODP Leg 113. In P. F. Barker, J. P. Kennett, and the Leg 113 Shipboard Scientific Party, *Proceedings of the Ocean Drilling Program, Scientific Results* 113: 865–880.

Kennett, J. P. and L. D. Stott. 1991. Abrupt deep-sea warming, paleoceanographic changes and benthic extinctions at the end of the Palaeocene. *Nature* 353: 225–229.

Kennett, J. P., C. von der Borch, P. A. Baker, C. E. Barton, A. Boersma, J. P. Cauler, W. C. Dudley, Jr., J. V. Gardner, D. G. Jenkins, W. H. Lohman, E. Martini, R. B. Merrill, R. Morin, C. S. Nelson, C. Robert, M. S. Srinivasan, R. Stein, A. Takeuchi, and M. G. Murphy. 1985. Paleotectonic implications of increased late Eocene–early Oligocene volcanism from South Pacific DSDP sites. *Nature* 316: 507–511.

Kitchell, J. A. and G. Estabrook. 1986. Was there a 26-Myr periodicity of extinctions? *Nature* 321: 534–535.

Kitchell, J. A. and D. Pena. 1984. Periodicity of extinctions in the geologic past: Deterministic versus stochastic explanations. *Science* 226: 689–692.

Koch, P. L., J. C. Zachos, and P. D. Gingerich. 1992. Correlation between isotope records in marine and continental carbon reservoirs near the Palaeocene/Eocene boundary. *Nature* 358: 319–322.

Krause, D. W. and M. C. Maas. 1990. The biogeographic origins of late Paleocene–early Eocene mammalian immigrants to the Western Interior of North America. *Geological Society of America Special Paper* 243: 71–105.

Krishtalka, L., R. K. Stucky, R. M. West, M. C. McKenna, C. C. Black, T. M. Bown, M. R. Dawson, D. J. Golz, J. J. Flynn, J. A. Lillegraven, and W. D. Turnbull. 1987. Eocene (Wasatchian through Duches-nean) biochronology of North America. In M. O. Woodburne, ed., *Cenozoic Mammals of North America, Geochronology and Biostratigraphy*, pp. 77–117. Berkeley: University of California Press.

Lanham, U. 1973. *The Bone Hunters*. New York: Columbia University Press.

Legendre, S. 1986. Analysis of mammalian communities from the late Eocène and Oligocene of southern France. *Palaeovertebrata* 16: 191–212.

Legendre, S. 1987. Concordance entre paléontologie continentale et les événements paléocéanographiques: exemple des faunes de mammifères du Paléogène du Quercy. *Comptes Rendus Hebdomadaires des Seances de l'Aca demie des Sciences (Sér. 3)* 304: 45–50.

Legendre, S. 1988. Le gisement du Bretou (Phosphorites du Quercy, Tannet-Garonne, France) et sa faune de vertebres de l'Eocene superieur. *Palaeontographica Abteilung A* 205: 173–182.

Legendre, S. 1989. Les communautés de mammifères du Paléogène (Eocène supérieur et Oligocène) d'Europe occidentale: Structures, milieux et évolution. *Münchner Geowissenschaften Abhandlung (A)* 16: 1–110.

Legendre, S. and J. L. Hartenberger. 1992. The evolution of mammalian faunas in Europe during the Eocene and Oligocene. In D. R. Prothero and W. A. Berggren, eds., *Eocene-Oligocene Climatic and Biotic Evolution*, pp. 516–528. Princeton: Princeton University Press.

Leidy, J. 1873. Contributions to the Extinct Vertebrate Fauna of the Western Territories. *Report of the United States Geological Survey of the Territories* 1.

REFERENCES

LeMasurier, W. E. 1972. Volcanic record of Cenozoic glacial history of Marie Byrd Land. In R. J. Adie, ed., *Antarctic Geology and Geophysics*, pp. 251–259. Oslo: Universitetsforlaget.

LeMasurier, W. E. and D. C. Rex. 1982. Volcanic record of Cenozoic glacial history in Marie Byrd Land and western Ellsworth Land: Revised chronology and evaluation of tectonic factors. In C. Craddock, ed., *Antarctic Geoscience*, pp. 725–734. Madison: University of Wisconsin Press.

Leopold, E. B., Liu Gengwu, and S. Clay-Poole. 1992. Low-biomass vegetation in the Oligocene? In D. R. Prothero and W. A. Berggren, eds., *Eocene-Oligocene Climatic and Biotic Evolution*, pp. 399–420. Princeton: Princeton University Press.

Li, C. K. and S. Y. Ting. 1983. The Paleogene mammals of China. *Bulletin of the Carnegie Museum of Natural History* 21: 1–93.

Lillegraven, J. A. 1972. Ordinal and familial diversity of Cenozoic mammals. *Taxon* 21: 261–274.

Lillegraven, J. A. 1979. A biogeographical problem involving compari-sons of late Eocene terrestrial vertebrate faunas of western North America. In J. Gray and A. J. Boucot, eds., *Historical Biogeog·raphy, Plate Tectonics, and the Changing Environment*, pp. 333–347. Corvallis, Oregon: Oregon State University Press.

Loper, D. E. and K. McCartney. 1986. Mantle plumes and the periodicity of magnetic field reversals. *Geophysical Research Letters* 13: 1525–1528.

Loper, D. E., K. McCartney, and G. Buzyna. 1988. A model of correlated episodicity in magnetic-field reversals, climate, and mass extinctions. *Journal of Geology* 96: 1–15.

Lopez, N. and L. Thaler. 1974. Sur les plus ancien lagomorphe européen et la "Grande Coupure" Oligocène de Stehlin. *Palaeovertebrata* 6: 243–251.

Loutit, T. S. and J. P. Kennett. 1981. New Zealand and Australian Cenozoic sedimentary cycles and global sea level changes. *American Association of Petroleum Geologists Bulletin* 65: 1586–1601.

Lowrie, W. and W. Alvarez. 1981. One hundred million years of geomagnetic polarity history. *Geology* 9: 392–397.

Lowrie, W., G. Napoleone, K. Perch-Nielsen, I. Premoli-Silva and M. Toumarkine. 1982. Paleogene magnetic stratigraphy in Umbrian pelagic carbonate rocks: The Contessa sections, Gubbio. *Geologi-cal Society of America Bulletin* 92: 414–432.

Lucas, S. G. 1992. Redefinition of the Duchesnean land mammal "age," late Eocene of western North America. In D. R. Prothero and W. A. Berggren, eds., *Eocene-Oligocene Climatic and Biotic Evolution*, pp. 88–105. Princeton: Princeton University Press.

Lucas, S. G. and R. M. Schoch. 1981. *Basalina*, a tillodont from the

Eocene of Pakistan. *Mitteilung Bayerische Staatsammlung Paläontologie und Historische Geologie* 21: 89–95.

Luckett, W.P. and J. L. Hartenberger. 1985. Evolutionary relationships among rodents: Comments and conclusions. In W. P. Luckett and J. L. Hartenberger, eds., *Evolutionary Relationships among Rodents; a multidisciplinary analysis*, pp. 685–712. New York: Plenum Press.

Lyell, Charles. 1831–1833. *Principles of Geology* (1st ed., 3 vols.) London: John Murray.

MacFadden, B.J., K. E. Campbell, R. L. Cifelli, O. Siles, N. M. Johnson, C. W. Naeser, and P. K. Zeitler. 1985. Magnetic polarity stratigraphy and mammalian faunas of the Deseadean (late Oligocene–early Miocene) Salla Beds of northern Bolivia. *Journal of Geology* 93: 223–250.

MacGinitie, H. D. 1953. Fossil plants of the Florissant beds, Colorado. *Carnegie Institute of Washington Publication* 599: 1–198.

MacGinitie, H. D. 1969. The Eocene Green River flora of northwestern Colorado and northeastern Utah. *University of California Publications in Geological Science* 83: 1–202.

MacGinitie, H. D. 1974. An early middle Eocene flora from the Yellowstone-Absaroka province, northwestern Wind River Basin, Wyoming. *University of California Publications in Geological Science* 108: 1–103.

MacLeod, N. 1990. Effects of late Eocene impacts on planktic foraminifera. *Geological Society of America Special Paper* 247: 595–606.

Mader, B. J. 1989. The Brontotheriidae: A systematic revision and preliminary phylogeny of North American genera. In D. R. Prothero and R. M. Schoch, eds., *The Evolution of Perissodactyls*, pp. 458–484. New York: Oxford University Press.

Margolis, S. V. and J. P. Kennett. 1971. Cenozoic paleoglacial history of Antarctica recorded in subantarctic deep-sea cores. *American Journal of Science* 271: 1–36.

Marsh, O. C. 1873. On the gigantic fossil mammals of the order Dinocerata. *American Journal of Science* 3 (v): 117-122.

Marsh, O. C. 1886. Dinocerata: A monograph of an extinct order of gigantic mammals. *Monograph of the United States Geological Survey* 10.

Marsh, O. C. 1896. The dinosaurs of North America. *Annual Reports of the United States Geological Survey* 16(1): 133–244.

Marshall, L. G. 1985. Geochronology and land-mammal biochronology of the transamerican faunal interchange. In F. G. Stehli and S. D. Webb, eds., *The Great American Biotic Interchange*, pp. 49–88. New York: Plenum Press.

Marshall, L. G. and R. L. Cifelli. 1989. Analysis of changing diversity

REFERENCES

patterns in Cenozoic land mammal age faunas, South America. *Palaeovertebrata* 19: 169–210.

Martini, E. and S. Ritzkowski. 1968. Was ist das 'Unter-Oligocän?' *Akad. Wiss. Gottingen, Nachr. Math. Phys. Kl.* 13: 231–250.

Matthew, W. D. 1899. A provisional classification of the fresh-water Tertiary of the West. *Bulletin of the American Museum of Natural History* 12: 19–75.

Matthews, R. K. and R. Z. Poore. 1980. Tertiary $d^{18}O$ record and glacio-eustatic sea-level fluctuations. *Geology* 8: 501–504.

Maurrasse, F. and B. P. Glass. 1976. Radiolarian stratigraphy and North American microtektites in Caribbean Core RC9-58: Implications concerning late Eocene radiolarian chronology and the age of the Eocene-Oligocene boundary. *Seventh Caribbean Geological Conference Proceedings*, pp. 205–212.

Mayer-Eymar, C. 1893. Le Ligurien et le Tongrien en Egypte. *Bulletin Societe Géologique France* 3: 7–43.

McGowran, B. 1973. Observation Borehole No. 2, Gambier Embayment of the Otway Basin: Tertiary micropaleontology and stratigraphy. *South Australian Department of Mines and Mineral Resources Review* 135: 43–55.

McGowran, B. 1978. Stratigraphic record of early Tertiary oceanic and continental events in the Indian Ocean region. *Marine Geology* 26: 1–39.

McGowran, B. 1989. Silica burp in the Eocene ocean. *Geology* 17: 857–860.

McGowran, B. 1990. Fifty million years ago. *American Scientist* 78: 30–39.

McGowran, B., G. Moss, and A. Beecroft. 1992. Late Eocene and early Oligocene in southern Australia: Local neritic signals of global oceanic changes. In D. R. Prothero and W. A. Berggren, eds., *Eocene-Oligocene Climatic and Biotic Evolution*, pp. 178–201. Princeton: Princeton University Press.

McIntyre, A. and A. H. W. Bé. 1967. Modern Coccolithophoridae of the Atlantic Ocean. *Deep Sea Research* 14: 561–597.

McKenna, M. C. 1967. Classification, range and deployment of the prosimian primates. *Colloques Internationaux du Centre National de la Recherche Scientifique* 163: 603–610.

McKenna, M. C. 1980. Eocene paleolatitude, climate and mammals of Ellesmere Island. *Palaeogeography, Palaeoclimatology, Palaeo-ecology* 30: 349–362.

McKenna, M. C. 1983a. Holarctic landmass rearrangement, cosmic events, and Cenozoic terrestrial organisms. *Annals of the Missouri Botanical Garden* 70: 459–489.

McKenna, M. C. 1983b. Cenozoic paleogeography of North Atlantic land bridges. In M. H. P. Bott, S. Saxov, M. Talwani and J. Theide, eds., *Structure and Development of the Greenland-Scotland*

Ridge. NATO Conference Series IV, Marine Sciences 8: 351–400. New York: Plenum Press.

McKenna, M. C., Chow Minchen, Ting Suyin, and Luo Zhexi. 1989. *Radinskya yupingae*, a perissodactyl-like mammal from the late Paleocene of southern China. In D. R. Prothero and R. M. Schoch, eds., *The Evolution of Perissodactyls*, pp. 24–36. New York: Oxford University Press.

McKinney, M. L., B. D. Carter, K. J. McNamara and S. K. Donovan. 1992. Evolution of Paleogene echinoids: A global and regional view. In D. R. Prothero and W. A. Berggren, eds., *Eocene-Oligocene Climatic and Biotic Evolution*, pp. 348–367. Princeton: Princeton University Press.

Meek, F. B. and F. V. Hayden. 1857. Descriptions of new species and genera of fossils, collected by Dr. F. V. Hayden in Nebraska Territory. *Proceedings of the Academy of Natural Sciences of Philadelphia* 9: 117–148.

Mellett, J. S. 1982. Body size, diet, and scaling factors in large carnivores and herbivores. *Proceedings of the Third North American Paleontological Convention* 2: 371–376.

Meyer, H. W. 1986. An evaluation of the methods for estimating paleoaltitudes using Tertiary floras from the Rio Grande Rift valley, New Mexico and Colorado. Ph.D. thesis. University of California, Berkeley.

Miall, A. D. 1992. Exxon global cycle chart: An event for every occasion? *Geology* 20: 787–790.

Miller, K. G. 1992. Middle Eocene to Oligocene stable isotopes, climate, and deep-water history: The Terminal Eocene Event? In D. R. Prothero and W. A. Berggren, eds., *Eocene-Oligocene Climatic and Biotic Evolution*, pp. 160–177. Princeton: Princeton University Press.

Miller, K. G., W. A. Berggren, J. Zhang and A. A. Palmer-Julson. 1991. Biostratigraphy and isotope stratigraphy of upper Eocene microtektites at Site 612: How many impacts? *Palaios* 6: 17–38.

Miller, K. G. and W. B. Curry. 1982. Eocene to Oligocene benthic foraminiferal isotopic record of the Bay of Biscay. *Nature* 296: 347–350.

Miller, K. G. and R. G. Fairbanks. 1983. Evidence for Oligocene–Middle Miocene abyssal circulation changes in the western North Atlantic. *Nature* 306: 250–253.

Miller, K. G. and R. G. Fairbanks. 1985. Oligocene to Miocene global carbon isotope cycles and abyssal circulation changes. In E. T. Sundquist and W. S. Broecker, eds., *The Carbon Cycle and Atmospheric CO_2: Natural Variations Archean to Present: American Geophysical Union, Geophysical Monograph* 32: 469–486.

Miller, K. G., R. G. Fairbanks and G. S. Mountain. 1987. Tertiary oxygen isotope synthesis, sea level history, and continental margin erosion, *Paleoceanography* 2: 1–19.

Miller, K. G. and M. E. Katz. 1987. Oligocene to Miocene benthic foraminiferal and abyssal circulation changes in the North Atlantic. *Micropaleontology* 33: 97–149.

Miller, K. G. and E. Thomas. 1985. Late Eocene to Oligocene benthic foraminiferal isotopic record, Site 574, equatorial Pacific. *Initial Reports of the Deep Sea Drilling Project* 85: 771–777.

Miller, K. G. and B. E. Tucholke. 1983. Development of Cenozoic abyssal circulation south of the Greenland-Scotland Ridge. In M. H. P. Bott, S. Saxov, M. Talwani, and J. Thiede, eds., *Structure and Development of the Greenland-Scotland Ridge*, pp. 549–589. New York: Plenum Press.

Miller, K. G., J. D. Wright and R. G. Fairbanks. 1991. Unlocking the Ice House: Oligocene-Miocene oxygen isotopes, eustasy, and margin erosion. *Journal of Geophysical Research* 96: 6829–6848.

Mohr, B. A. R. 1990. Eocene and Oligocene sporomorphs and dinoflagellate cysts from Leg 113 drill sites, Weddell Sea, Antarctica. *Proceedings of the Ocean Drilling Program* 113: 595–606.

Montanari, A. 1990. Geochronology of the terminal Eocene impacts; an update. *Geological Society of America Special Paper* 247: 607–616.

Montanari, A., A. L. Deino, R. E. Drake, B. D. Turrin, D. J. DePaolo, G. S. Odin, G. H. Curtis, W. Alvarez and D. Bice. 1988. Radioisotopic dating of the Eocene-Oligocene boundary in the pelagic sequences of the northeastern Apennines. In I. Premoli-Silva, R. Coccioni, and A. Montanari, eds., *The Eocene-Oligocene Boun-dary in the Marche-Umbria Basin (Italy)*, pp. 195–208. Special Publication of the International Subcommission on Paleogene Stratigraphy, Eocene/Oligocene Boundary Meeting, Ancona, 1987.

Montanari, A., R. Drake, D. M. Bice, W. Alvarez, G. H. Curtis, B. Turrin, and D. J. DePaolo. 1985. Radiometric time scale for the upper Eocene and Oligocene based on K/Ar and Rb/Sr dating of volcanic biotites. *Geology* 13: 596–599.

Moore, T. C., T. H. van Andel, C. Sancetta, and N. Pisias. 1978. Cenozoic hiatuses in pelagic sediments. *Micropaleontology* 24: 113–138.

Mountain, G. S. and B. E. Tucholke. 1985. Mesozoic and Cenozoic geology of the U.S. Atlantic continental slope and rise. In C. W. Poag, ed., *Geologic Evolution of the United States Atlantic Margin*, pp. 292–341. New York: Van Nostrand Reinhold.

Munier-Chalmas, M. and A. de Lapparent. 1893. Note sur la nomenclature des terrains sédimentaires. *Bulletin Societé Géologique du France* (Série 3) 21: 438–488.

Murphy, M. G. and J. P. Kennett. 1986. Development of latitudinal thermal gradients during the Oligocene: Oxygen-isotope evidence from the southwest Pacific. *Initial Reports of the Deep Sea Drilling Project* 90: 1347–1360.

Mutter, J. C., K. A. Hegarty, S. C. Cande, and J. K. Weissel. 1985. Breakup between Australia and Antarctica: A brief review in light of new data. *Tectonophysics* 114: 255–279.

Ness, G., S. Levi, and R. Crouch. 1980. Marine magnetic anomaly timescales for the Cenozoic and Late Cretaceous: A precís, critique, and synthesis. *Reviews of Geophysics and Space Physics* 18(4): 753–770.

Nocchi, M., S. Monechi, R. Coccioni, M. Madile, P. Monaco, M. Orlando, G. Parisi and I. Premoli-Silva. 1988. The extinction of Hantkeninidae as a means for recognizing the Eocene-Oligocene boundary: A proposal. In I. Premoli-Silva, R. Coccioni, and A. Montanari, eds., *The Eocene-Oligocene Boundary in the Marche-Umbria Basin*, pp. 249–252. Special Publication of the International Subcommission on Paleogene Stratigraphy, Eocene/Oligocene Boundary Meeting, Ancona, 1987.

Nocchi, M., G. Parisi, P. Monaco, S. Monechi, M. Mandile, G. Napoleone, M. Ripepe, M. Orlando, I. Premoli-Silva and D. M. Bice. 1986. The Eocene–Oligocene boundary in the Cambrian pelagic regression. In C. Pomerol and I. Premoli-Silva, eds., *Terminal Eocene Events*, pp. 25–40. Amsterdam: Elsevier Science Publishers.

Noma, E. and A. L. Glass. 1987. Mass extinction pattern: Result of chance. *Geological Magazine* 124: 319–322.

Novacek, M. J. 1985. Cranial evidence for rodent affinities. In W. P. Luckett and J. L. Hartenberger, eds., *Evolutionary Relationships among Rodents; a multidisciplinary analysis*, pp. 59–81. New York: Plenum Press.

Novacek, M. J., A. Wyss, D. Frassinetti and P. Salinas. 1989. A new? Eocene mammal fauna from the Andean main range. *Journal of Vertebrate Paleontology* 9(3): 34A (abstract).

Obradovich, J. D. 1988. A different perspective on glauconite as a chronometer for geologic time-scale studies. *Paleoceanography* 3: 757–770.

Obradovich, J. D., G. A. Izett and C. W. Naeser. 1973. Radiometric ages of volcanic ash and pumice beds in the Gering Sandstone (earliest Miocene) of the Arikaree Group, southwestern Nebraska. *Geological Society of America, Abstracts with Programs* 5: 499–500.

Obradovich, J. D., L. W. Snee, and G. A. Izett. 1989. Is there more than one glassy impact layer in the late Eocene? *Geological Society of America Abstracts with Programs*, A134.

Odin, G. S. 1978. Isotopic dates for the Paleogene time scale.

REFERENCES

American Association of Petroleum Geologists Studies in Geology 6: 247–257.

Odin, G. S., ed. 1982. *Numerical Dating in Stratigraphy.* New York: John Wiley

Odin, G. S. and D. Curry. 1985. The Paleogene time scale: Radiometric dating versus magnetostratigraphic approach. *Journal of the Geological Society of London* 142: 1179–1188.

Odin, G. S. and A. Montanari. 1988. The Eocene-Oligocene boundary at Massignano (Ancona, Italy): A potential boundary stratotype. In I. Premoli-Silva, R. Coccioni and A. Montanari, eds., *The Eocene-Oligocene Boundary in the Marche-Umbria Basin,* pp. 253–263. Special Publication of the International Subcommission on Paleogene Stratigraphy, Eocene/Oligocene Boundary Meeting, Ancona, 1987.

Officer, C. B., A. Hallam, C. L. Drake, and J. D. Devine. 1987. Late Cretaceous and paroxysmal Cretaceous/Tertiary extinctions. *Nature* 326: 143–149.

Olsson, R. K., K. G. Miller, and T. E. Ungrady. 1980. Late Oligocene transgression of middle Atlantic coastal plain. *Geology* 8: 549–554.

Osborn, H. F. 1907. Tertiary mammal horizons of North America. *Bulletin of the American Museum of Natural History* 23: 237–254.

Osborn, H. F. 1910. *The Age of Mammals in Europe, Asia, and North America.* New York: MacMillan.

Osborn, H. F. 1929. The titanotheres of ancient Wyoming, Dakota, and Nebraska. *United States Geological Survey Monograph* 55: 1–953 (2 vols.).

Osborn, H. F. 1930. *Fifty-two Years of Research, Observation and Publication.* New York: Scribner's.

Osborn, H. F. and W. D. Matthew. 1909. Cenozoic mammal horizons of western North America. *United States Geological Survey Bulletin* 361: 1–138.

Owen, R. M. and D. K. Rea. 1985. Sea-floor hydrothermal activity links climate to tectonics: The Eocene carbon dioxide greenhouse. *Science* 227: 166–169.

Owens, J. P., L. M. Bybell, G. Paulachok, T. A. Ager, V. M. Gonzalez, and P. J. Sugarman. 1988. Stratigraphy of the Tertiary sediments in a 945-foot-deep corehole near Mays Landing in the southeastern New Jersey Coastal Plain. *United States Geological Survey Professional Paper* 1484.

Parrish, J. T. 1987. Global paleogeography and paleoclimate of the late Cretaceous and early Tertiary. In E. M. Friis, W. G. Chaloner, and P. R. Crane, eds., *The Origin of Angiosperms and their Biological Consequences,* pp. 51–73. Cambridge: Cambridge University Press.

Pascual, R. and E. Ortiz Jaureguizar. 1990. Evolving climates and mammal faunas in Cenozoic South America. *Journal of Human Evolution* 19: 23–60.

Pascual, R., M. G. Vucetich, G. J. Scillato-Yané, and M. Bond. 1985. Main pathways of mammalian diversification in South America. In F. G. Stehli and S. D. Webb, eds., *The Great American Biotic Interchange*, pp. 219–248. New York: Plenum Press.

Patterson, C. and A. B. Smith. 1987. Is the periodicity of extinctions a taxonomic artefact? *Nature* 330: 248–251.

Peterson, G. L. and P. L. Abbott. 1979. Mid-Eocene climatic change, southwestern California and northwestern Baja California. *Palaeogeography, Palaeoclimatology, Palaeoecology* 26: 73–87.

Plint, A. G. 1988. Global eustacy and the Eocene sequence in the Hampshire Basin, England. *Basin Res.* 1: 11–22.

Poag, C. W., D. S. Powars, L. J. Poppe, R. B. Mixon, L. E. Edwards, D. W. Folger, and S. Bruce. 1992. Deep Sea Drilling Project Site 612 bolide event: New evidence of late Eocene impact-wave deposits and a possible impact site, U.S. east coast. *Geology* 20: 771–774.

Poag, C. W. and J. S. Schlee. 1984. Depositional sequences and stratigraphic gaps on submerged United States Atlantic margin. *American Association of Petroleum Geologists Memoir* 36: 165–182.

Poag, C. W. and L. W. Ward. 1987. Cenozoic unconformities and depositional supersequences of North Atlantic continental margins: Testing the Vail model. *Geology* 15: 159–162.

Pomerol, C. and I. Premoli-Silva, eds., 1986. *Terminal Eocene Events*. Amsterdam: Elsevier.

Poore, R. Z., L. Tauxe, S. F. Percival, Jr. and J. L. LaBrecque. 1982. Late Eocene–Oligocene magnetostratigraphy and biostratigraphy at South Atlantic DSDP Site 527. *Geology* 10: 508–511.

Premoli-Silva, I., R. Coccioni, and A. Montanari (eds.). 1988. *The Eocene/Oligocene Boundary in the Marche-Umbria Basin (Italy)*. Special Publication of the International Subcommission on Paleogene Stratigraphy, Eocene/Oligocene Boundary Meeting, Ancona, 1987.

Prentice, M. L. and R. K Matthews. 1988. Cenozoic ice-volume history: Development of a composite oxygen isotope record. *Geology* 16: 963–966.

Prothero, D. R. 1985a. Chadronian (early Oligocene) magnetostratigraphy of eastern Wyoming: Implications for the Eocene-Oligocene boundary. *Journal of Geology* 93: 555–565.

Prothero, D. R. 1985b. Mid-Oligocene extinction event in North American land mammals. *Science* 229: 550–551.

Prothero, D. R. 1985c. Correlation of the White River Group by magnetostratigraphy. In J. E. Martin, ed., Fossiliferous Cenozoic deposits of western South Dakota and northwestern Nebraska.

REFERENCES

Dakoterra 2(2): 265–276. Museum of Geology, South Dakota School of Mines.

Prothero, D. R. 1985d. North American mammalian diversity and Eocene-Oligocene extinctions. *Paleobiology* 11(4): 389–405.

Prothero, D. R. 1989. Stepwise extinctions and climatic decline during the later Eocene and Oligocene. In S. K. Donovan, ed., *Mass Extinctions: Processes and Evidence*, pp. 211–234. New York: Columbia University Press.

Prothero, D. R. 1990. *Interpreting the Stratigraphic Record*. New York: W. H. Freeman.

Prothero, D. R. and J. M. Armentrout. 1985. Magnetostratigraphic correlation of the Lincoln Creek Formation, Washington: Implications for the age of the Eocene-Oligocene boundary. *Geology* 13: 208–211.

Prothero, D. R. and W. A. Berggren. 1992. *Eocene-Oligocene Climatic and Biotic Evolution*. Princeton: Princeton University Press.

Prothero, D. R., W. A. Berggren, and P. R. Bjork. 1990. Penrose Conference Report: Late Eocene–Oligocene biotic and climatic evolution. *Geological Society of America News and Information* 12(3): 74–75.

Prothero, D. R., C. R. Denham, and H. G. Farmer. 1982. Oligocene calibration of the magnetic polarity time scale. *Geology* 10: 650–653.

Prothero, D. R., C. R. Denham, and H. G. Farmer. 1983. Magnetostratigraphy of the White River Group and its implications for Oligocene geochronology. *Palaeogeography, Palaeoclimatology, Palaeoecology* 42: 151–166.

Prothero, D. R. and R. J. Emry, eds. 1995. *The Terrestrial Eocene-Oligocene Transition in North America*. Cambridge: Cambridge University Press.

Prothero, D. R., C. Guérin, and E. Manning. 1989. The history of the Rhinocerotoidea. In D. R. Prothero and R. M. Schoch, eds., *The Evolution of Perissodactyls*, pp. 322–340. New York: Oxford University Press.

Prothero, D. R., E. Manning, and C. B. Hanson. 1986. The phylogeny of the Rhinocerotoidea (Mammalia, Perissodactyla). *Zoological Journal of the Linnean Society of London* 87: 341–366.

Prothero, D. R. and R. M. Schoch. 1989. Origin and evolution of the Perissodactyla: A summary and synthesis. In D. R. Prothero and R. M. Schoch, eds., *The Evolution of Perissodactyls*, pp. 504–529. New York: Oxford University Press.

Prothero, D. R. and R. M. Schoch. 1995. *Horns, Tusks, Hooves, and Flippers: The Evolution of Hoofed Mammals and their Relatives.* Princeton: Princeton University Press.

Prothero, D. R. and C. C. Swisher III. 1992. Magnetostratigraphy and geochronology of the terrestrial Eocene-Oligocene transition in North America. In D. R. Prothero and W. A. Berggren, eds., *Eocene-Oligocene Climatic and Biotic Evolution*, pp. 46–73. Princeton: Princeton University Press.

Quilty, P. G. 1977. Cenozoic sedimentation cycles in western Australia. *Geology* 5: 336–340.

Quinn, J. F. 1987. On the statistical detection of cycles in extinctions in the marine fossil record. *Paleobiology* 13: 456–478.

Rainger, R. 1991. *An Agenda for Antiquity: Henry Fairfield Osborn and Vertebrate Paleontology at the American Museum of Natural History, 1890–1935*. Tuscaloosa, Alabama: University of Alabama Press.

Rampino, M. R. and R. B. Stothers. 1984. Terrestrial mass extinctions, cometary impacts, and the Sun's motion perpendicular to the galactic plane. *Nature* 308: 709–712.

Rampino, M. R. and R. B. Stothers. 1988. Flood basalt volcanism during the past 250 million years. *Science* 241: 663–668.

Rasmussen, D. T., T.M. Bown, and E. L. Simons. The Eocene-Oligocene transition in continental Africa. In D. R. Prothero and W. A. Berggren, eds., *Eocene-Oligocene Climatic and Biotic Evolution*, pp. 548–566. Princeton: Princeton University Press.

Raup, D. M. 1985. *The Nemesis Affair: A Story of the Death of the Dinosaurs and the Ways of Science*. New York: W. W. Norton.

Raup, D. M. 1991. *Extinction: Bad Genes or Bad Luck?* New York: W. W. Norton.

Raup, D. M. and J. J. Sepkoski, Jr. 1982. Mass extinctions in the marine fossil record. *Science* 215: 1501–1503.

Raup, D. M. and J. J. Sepkoski, Jr. 1984. Periodicity of extinctions in the geologic past. *Proceedings of the National Academy of Sciences* 81: 805–801.

Raup, D. M. and J. J. Sepkoski, Jr. 1986. Periodic extinctions of families and genera. *Science* 231: 833–836.

Rea, D. K., J. C. Zachos, R. M. Owen, and P. D. Gingerich. 1990. Global change at the Paleocene-Eocene boundary: Climatic and evolutionary consequences of tectonic events. *Palaeogeography, Palaeoclimatology, Palaeoecology* 79: 117–128.

Retallack, G. J. 1981. Preliminary observations on fossil soils in the Clarno Formation (Eocene to early Oligocene), near Clarno, Oregon. *Oregon Geology* 43: 147–150.

Retallack, G. J. 1983a. Late Eocene and Oligocene paleosols from Badlands National Park, South Dakota. *Geological Society of America Special Paper* 193.

Retallack, G. J. 1983b. A paleopedological approach to the interpretation of terrestrial sedimentary rocks: The mid-Tertiary fossil soils of Badlands National Park, South Dakota. *Geological Society of America Bulletin* 94: 823–840.

Retallack, G. J. 1990. *Soils of the past: An introduction to paleo-*

REFERENCES

pedology. London: Unwin–Hyman.

Retallack, G. J. 1992. Paleosols and changes in climate and vegetation across the Eocene/Oligocene boundary. In D. R. Prothero and W. A. Berggren, eds., *Eocene-Oligocene Climatic and Biotic Evolu tion*, pp. 382–398. Princeton: Princeton University Press.

Ricou, L., B. Mercier de Lepinay, and J. Marcoux. 1986. Evolution of the Tethyan seaways and implications for oceanic circulation around the Eocene/Oligocene boundary. In C. Pomerol and I. Premoli-Silva, eds., *Terminal Eocene Events*, pp. 388–395. Amsterdam: Elsevier.

Ritzkowski, S. 1981. Latdorfian. *Bulletin Informationale Géologie des Bassins du Paris, Mémoire hors* Série 2: 149–166.

Roberts, D. G., A. C. Morton and J. Backmann. 1984. Late Paleocene–Eocene volcanic events in the northern North Atlantic Ocean. *Initial Reports of the Deep Sea Drilling Project* 81: 913–923.

Rona, P. A. and E. A. Richardson. 1978. Early Cenozoic global plate reorganization. *Earth and Planetary Science Letters* 40: 1–11.

Rooth, C. 1982. Hydrology and ocean circulation. *Progress in Oceanography* 11: 131–149.

Rudwick, M. J. S. 1978. Charles Lyell's dream of a statistical paleontology. *Palaeontology* 21: 225–244.

Russell, D. E. and R. Zhai. 1987. The Paleogene of Asia: Mammals and stratigraphy. *Memoirs Musee National d'Histoire Naturelle* (C) 51: 1–488.

Sanborn, E. I. 1935. The Comstock flora of west central Oregon. *Carnegie Institute of Washington Publication* 465: 1–28.

Sanborn, E. I. 1947. The Scio flora of western Oregon. *Oregon State Monograph, Studies in Geology* 4: 1–29.

Sanfilippo, A., W. R. Riedel, B. P. Glass and F. T. Kyte. 1985. Late Eocene microtektites and radiolarian extinctions on Barbados. *Nature* 314: 613–615.

Savage, D. E. and D. E. Russell. 1983. *Mammalian Paleofaunas of the World*. Reading, Massachusetts: Addison Wesley.

Savin, S. M. 1977. The history of the earth's surface temperature during the past 100 million years. *Annual Reviews of Earth and Planetary Sciences* 5: 319–355.

Savin, S. M., R. G. Douglas, and F. G. Stehli. 1975. Tertiary marine paleotemperatures. *Geological Society of America Bulletin* 86: 1499–1510.

Schaal, S. and W. Ziegler, eds., 1992. *Messel, an insight into the history of life and of the earth*. Oxford: Clarendon Press.

Schlich, R. and the Leg 120 Shipboard Scientific Party. 1989. *Proceedings of the Ocean Drilling Project, Initial Reports* 120.

Schoch, R. M. 1989a. A review of the tapiroids. In D. R. Prothero and

R. M. Schoch, eds., *The Evolution of Perissodactyls*, pp. 398–321. New York: Oxford University Press.

Schoch, R. M. 1989b. *Stratigraphy, Principles and Methods*. New York: Van Nostrand Reinhold.

Schultz, C. B. and C. H. Falkenbach. 1968. The phylogeny of the oreodonts, parts 1 and 2. *Bulletin of the American Museum of Natural History* 139: 1–148.

Schwartz, R. D. and P. B. James. 1984. Periodic mass extinctions and the Sun's oscillation around the galactic plane. *Nature* 308: 712–713.

Schweitzer, H. J. 1980. Environment and climate in the early Tertiary of Spitsbergen. *Palaeogeography, Palaeoclimatology, Palaeoecology* 30: 297–311.

Sclater, J. G., L. Meinke, A. Bennett, and C. Murphy. 1986. The depth of the ocean through the Neogene. *Geological Society of America Memoir* 163: 1–19.

Scott, W. B. 1945. The Mammalia of the Duchesne River Oligocene. *Transactions of the American Philosophical Society* 34: 209–253.

Scott, W. B., G. L. Jepsen, and A. E. Wood. 1940–1941. The mammalian fauna of the White River Oligocene. *Transactions of the American Philosophical Society* 28, parts I–V.

Sepkoski, J. J., Jr. 1989. Periodicity in extinction and the problem of catastrophism in the history of life. *Journal of the Geological Society of London* 146: 7–19.

Shackleton, N. J. 1986. Paleogene stable isotope events. *Palaeogeography, Palaeoclimatology, Palaeoecology* 57: 91–102.

Shackleton, N. J. and J. P. Kennett. 1975. Paleotemperature history of the Cenozoic and initiation of Antarctic glaciation: Oxygen and carbon isotopic analyses in DSDP Sites 277, 279, and 281. *Initial Reports of the Deep Sea Drilling Project* 29: 743–755.

Shackleton, N. J. and N. D. Opdyke. 1973. Oxygen isotope and paleomagnetic stratigraphy of Equatorial Pacific core V28-238: Oxygen isotope temperatures and ice volumes on a 10^5 year and 10^6 year scale. *Quaternary Research* 3: 39–55.

Shoemaker, E. M. and R. F. Wolfe. 1986. Mass extinctions, crater ages, and comet showers. In R. S. Smoluchowski, J. N. Bahcall, and M. S. Matthews, eds., *The Galaxy and the Solar System*, pp. 338–386. Tucson: University of Arizona Press.

Siesser, W. G. and R. V. Dingle. 1981. Tertiary sea-level movements around southern Africa. *Journal of Geology* 89: 83–96.

Sigé, B. and M. Vianey-Liaud. 1979. Impropriété de la Grande Coupure de Stehlin comme support d'une limite Eocène-Oligocène. *Newsletters in Stratigraphy* 8: 79–82.

Simpson, G. G. 1946. The Duchesnean fauna and the Eocene-Oligocene boundary. *American Journal of Science* 244: 52–57.

Sloan, L. C., J. C. G. Walker, T. C. Moore, Jr., D. K. Rea, and J. C.

REFERENCES

Zachos. 1992. Possible methane-induced polar warming in the early Eocene. *Nature* 357: 320–322.

Sloan, R. E. 1969. Cretaceous and Paleocene terrestrial mammal communities of western North America. *Proceedings of the North American Paleontological Convention* 1(E): 427–453.

Sloan, R. E. 1987. Paleocene and latest Cretaceous mammal ages, biozones, magnetozones, rates of sedimentation, and evolution. *Geological Society of America Special Paper* 209: 165–200.

Smith, A. B. and C. Patterson. 1988. The influence of taxonomic method on the perceptions of patterns of evolution. *Evolutionary Biology* 23: 127–216.

Snyder, S. W., C. Muller ,and K. G. Miller. 1984. Biostratigraphy and paleoceanography across the Eocene/Oligocene boundary at Site 549. *Geology* 12: 112–115.

Stanley, S. M. 1986. *Extinction.* New York: Scientific American.

Stanley, S. M. 1990. Delayed recovery and the spacing of major extinctions. *Paleobiology* 16: 401–414.

Stanley, S. M., W. O. Addicott, and K. Chinzei. 1980. Lyellian curves in paleontology: possibilities and limitations. *Geology* 8: 422–426.

Stehlin, H. G. 1909. Remarques sur les faunules de mammifères des couches éocènes et oligocènes du Bassin de Paris. *Bulletin Societeé Géologique de France* 9: 488–520.

Stehlin, H. G. 1910. Die Säugetiere des schweizerischen Eocaens 6. *Abh. Schweiz. Pal. Ges.* 26: 839–1164.

Stigler, S. M. and M. J. Wagner. 1987. A substantial bias in nonparametric tests for periodicity in geophysical data. *Science* 238: 940–945.

Stott, L. D. 1992. Higher temperatures and lower oceanic pCO_2: A climate enigma at the end of the Paleocene Epoch. *Paleoceanography* 7: 395–404.

Stucky, R. K. 1990. Evolution of land mammal diversity in North America during the Cenozoic. *Current Mammalogy* 2: 375–432.

Stucky, R. K. 1992. Mammalian faunas in North America of Bridgerian to early Arikareean "ages" (Eocene and Oligocene). In D. R. Prothero and W. A. Berggren, eds., *Eocene-Oligocene Climatic and Biotic Evolution*, pp. 464–493. Princeton: Princeton University Press.

Swisher, C. C., III, and D. R. Prothero. 1990. Single-crystal $^{40}Ar/^{39}Ar$ dating of the Eocene-Oligocene transition in North America. *Science* 249: 760–762.

Szalay, F. S. and Li Chuankei. 1986. Middle Paleocene euprimate from southern China and the distribution of primates in the Paleocene. *Journal of Human Evolution* 15: 387–397.

Talwani, M. and O. Eldholm. 1977. Evolution of the Norwegian-Greenland Sea. *Geological Society of America Bulletin* 88: 969–999.

REFERENCES

Taylor, S. R. 1973. Tektites: A post-Apollo view. *Earth Science Reviews* 101–103.

Tedford, R. H., T. Galusha, M. F. Skinner, B. E. Taylor, R. W. Fields, J. R. Macdonald, J. M. Rensberger, S. D. Webb, and D. P. Whistler. 1987. Faunal succession and biochronology of the Arikareean through Hemphillian interval (late Oligocene through earliest Pliocene Epochs) in North America. In M. O. Woodburne, ed., *Cenozoic Mammals of North America, Geochronology and Biostratigraphy*, pp. 152–210. Berkeley: University of California Press.

Tedford, R. H., M. R. Banks, N. R. Kemp, I. McDougall and F. L. Sutherland. 1975. Recognition of the oldest known fossil marsupials from Australia. *Nature* 255: 141–142.

Thewissen, J. G. M., D. E. Russell, P. D. Gingerich and S. T. Hussain. 1983. A new dichobunid artiodactyl (Mammalia) from the Eocene of northwest Pakistan. *Proceedings of the Koninklikje Nederlandse Akademie van Wetenschappen (B)* 86: 153–180.

Thomas, E. 1992. Middle Eocene–late Oligocene bathyal benthic foraminifera (Weddell Sea): Faunal changes and implications for oceanic circulation. In D. R. Prothero and W. A. Berggren, eds., *Eocene-Oligocene Climatic and Biotic Evolution*, pp. 245–271. Princeton: Princeton University Press.

Thomson, K. S. 1988. Anatomy of the extinction debate. *American Scientist* 76: 59–61.

Tremaine, S. D. 1986. Is there evidence of a solar companion star? In R. S. Smoluchowski, J. N. Bahcall and M.S. Matthews, eds., *The Galaxy and the Solar System*, pp. 409–416. Tucson: University of Arizona Press.

Truswell, E. M. 1983. Recycled Cretaceous and Tertiary pollen and spores in Antarctic marine sediments: A catalogue. *Palaeontographica (B)* 186: 121–174.

Truswell, E. M. and W. K. Harris. 1982. The Cainozoic palaeobotanical record in arid Australia: Fossil evidence for the origins of arid-adapted flora. In W. R. Barker and P. J. M. Greenslade, eds., *Evolution of the Flora and Fauna of Arid Australia*, pp. 67–76. Adelaide: Peacock Publications.

Vail, P. R., R. M. Mitchum, Jr., and S. Thompson III. 1977. Global cycles of relative changes in sea level. *American Association of Petroleum Geologists Memoir* 26: 83–98.

Vanyo, J. P. and S. M. Aramwik. 1982. Length of day obliquity at the ecliptic 850 Ma ago: Preliminary results of a stromatolitic growth model. *Geophysical Research Letters* 9: 1125–1128.

Veevers, J. J. 1986. Breakup of Australia and Antarctica estimated as mid-Cretaceous (95 ± 5 Ma) from magnetic and seismic data at the continental margin. *Earth and Planetary Science Letters* 77: 91–99.

REFERENCES

Vianey-Liaud, M. 1976. *L'évolution des Rongeurs à l'Oligocène en Europe occidentale.* Montpellier, France: Ph. D. Thèse.

Vianey-Liaud, M. 1991. Les rongeurs de l'Eocène terminal et de l'Oligocène d'Europe comme indicateurs de leur environment. *Palaeogeography, Palaeoclimatology, Palaeoecology* 85: 15–28.

Vincent, E. and W. H. Berger. 1985. Carbon dioxide and polar cooling in the Miocene: The Monterey hypothesis. In E. T. Sundquist and W. S. Broecker, eds., The Carbon Cycle and Atmospheric CO_2: Natural Variations, Archean to Present. *American Geophysical Union Geophysical Monograph* 32: 455–468.

Wang Banyue. 1992. The Chinese Oligocene: A preliminary review of mammalian localities and local faunas. In D. R. Prothero and W. A. Berggren, eds., *Eocene-Oligocene Climatic and Biotic Evolu tion*, pp. 529–547. Princeton: Princeton University Press.

Ward, W. R. 1982. Comments on the long-term stability of the Earth's obliquity. *Icarus* 50: 444–448.

Warren, B. A. 1971. Antarctic deep-water circulation contribution to the world ocean. In L. O. Quam, ed., *Research in the Antarctic* 93: 640–643. Washington, D.C.: American Association for the Advancement of Science.

Watkins, J. S. and G. S. Mountain, eds. 1990. *Role of ODP Drilling in the Investigation of Global Change in Sea Level.* Report of a JOI/USSAC Workshop, El Paso, Texas, October 1988. (Unpublished report, available from JOI/USSAC Office, Washington, D.C.)

Webb, P. N., D. M. Harwood, B. C. McKelvey, J. H. Mercer and L. D. Stott. 1984. Cenozoic marine sedimentation and ice-volume variation on the East Antarctic craton. *Geology* 12: 287–291.

Webb, S. D. 1977. A history of savanna vertebrates in the New World; Part I: North America. *Annual Reviews of Ecology and Systematics* 8: 355–380.

Webb, S. D. 1983. The rise and fall of the late Miocene ungulate fauna in North America. In M. D. Nitecki, ed., *Coevolution*, pp. 267–306. Chicago: University of Chicago Press.

Wei Wuchang. 1989. Reevaluation of the Eocene ice-rafting record from subantarctic cores. *Antarctic Journal of the United States* 1989: 108–109.

Weissel, J. K., D. E. Hayes and E. M. Herron. 1977. Plate tectonics synthesis: The displacements between Australia, New Zealand, and Antarctica since the late Cretaceous. *Marine Geology* 25: 231–277.

Whitmire, D. P. and A. A. Jackson IV. 1984. Are periodic mass extinctions driven by a distant solar companion? *Nature* 308: 713–715.

Whitmire, D. P. and A. A. Jackson IV. 1985. Periodic comet showers and Planet X. *Nature* 313: 36–38.

Wilgus, C. K., B. S. Hastings, C. G. St. C. Kendall, H. W. Posamentier, C. A. Ross, and J. C. Van Wagoner, eds. 1988. *Sea-Level Changes: An Integrated Approach. S. E. P. M. Special Publication* 42.

Williams, C. A. 1986. An oceanwide view of Palaeogene plate tectonic events. *Palaeogeography, Palaeoclimatology, Palaeoecology* 57: 3–25.

Wilson, J. A. 1978. Stratigraphic occurrence and correlation of early Tertiary vertebrate faunas, Trans-Pecos Texas; Part 1: Vieja area. *Texas Memorial Museum Bulletin* 25: 1–42.

Wilson, J. A. 1984. Vertebrate fossil faunas 49 to 36 million years ago and additions to the species of *Leptoreodon* found in Texas. *Journal of Vertebrate Paleontology* 4: 199–207.

Wilson, J. A. 1986. Stratigraphic occurrence and correlation of early Tertiary vertebrate faunas, Trans-Pecos Texas: Agua Fria–Green Valley areas. *Journal of Vertebrate Paleontology* 6: 350–373.

Wilson, R. W. 1972. Evolution and extinction in early Tertiary rodents. *Proceedings of the 24th International Geological Congress* 7: 217–222.

Wing, S. L. 1987. Eocene and Oligocene floras and vegetation of the Rocky Mountains. *Annals of the Missouri Botanical Garden* 74: 748–784.

Wise, S. W. Jr., J. R. Breza, D. M. Harwood, W. Wei, and J. C. Zachos. 1992. Paleogene glacial history of Antarctica in light of Leg 120 drilling results. *Proceedings of the Ocean Drilling Program, Scientific Results* 120: 1001-1028.

Wolfe, J. A. 1969. *Paleogene flora from the Gulf of Alaska region. United States Geological Survey Open-File Report.*

Wolfe, J. A. 1971. Tertiary climatic fluctuations and methods of analysis of Tertiary floras. *Palaeogeography, Palaeoclimatology, Palaeoecology* 9: 27–57.

Wolfe, J. A. 1978. A paleobotanical interpretation of Tertiary climates in the Northern Hemisphere. *American Scientist* 66: 694–703.

Wolfe, J. A. 1980. Tertiary climates and floristic relationships at high latitudes in the Northern Hemisphere. *Palaeogeography, Palaeoclimatology, Palaeoecology* 30: 313–323.

Wolfe, J. A. 1985. Distributions of major vegetational types during the Tertiary. In E. T. Sundquist and W. S. Broecker, eds., The carbon cycle and atmospheric CO_2: Natural variations, Archean to present. *American Geophysical Union Geophysical Monographs* 32: 357–376.

Wolfe, J. A. 1986. Tertiary floras and paleoclimates of the Northern Hemisphere. In T. W. Broadhead, ed., Land Plants: Notes for a short course. *University of Tennessee Department of Geological*

REFERENCES

Sciences Studies in Geology 15: 182–196.

Wolfe, J. A. 1990. Estimates of Pliocene precipitation and temperature based on multivariate analysis of leaf physiognomy. *United States Geological Survey Open-File Report* 90-94: 39–42.

Wolfe, J. A. 1992. Climatic, floristic, and vegetational changes near the Eocene/Oligocene boundary in North America. In D. R. Prothero and W. A. Berggren, eds., *Eocene-Oligocene Climatic and Biotic Evolution*, pp. 421–436. Princeton: Princeton University Press.

Wolfe, J. A. and D. M. Hopkins. 1967. Climatic changes recorded by Tertiary land floras in northwestern North America. In K. Hatai, ed., *Tertiary Correlation and Climatic Changes in the Pacific*, pp. 67–76. Tokyo: Sasaki Printing and Publishing.

Wolfe, J. A. and R. Z. Poore. 1982. Tertiary marine and nonmarine climatic trends. In W. Berger and J. C. Crowell, eds., *Climate in Earth History*, pp. 154–158. Washington, D.C.: National Academy of Sciences.

Wood, H. E., II, R. W. Chaney, J. Clark, E. H. Colbert, G. L. Jepsen, J. B. Reeside, Jr., and C. Stock. 1941. Nomenclature and correlation of the North American continental Tertiary. *Geological Society of America Bulletin* 52: 1–48.

Woodburne, M. O., ed. 1987. *Cenozoic Mammals of North America: Geochronology and Biostratigraphy*. Berkeley: University of California Press.

Woodburne, M. O. and W. J. Zinsmeister. 1982. Fossil land mammal from Antarctica. *Science* 218: 284–286.

Woodburne, M. O. and W. J. Zinsmeister. 1984. The first land mammal from Antarctica and its biogeographic implications. *Journal of Paleontology* 58: 913–948.

Wyss, A. R., M. R. Norell, J. J. Flynn, M. J. Novacek, R. Charrier, M. C. McKenna, C. C. Swisher III, D. Frassinetti, P. Salinas, and Meng Jin. 1990. A new early Tertiary mammal fauna from central Chile: Implications for Andean stratigraphy and tectonics. *Journal of Vertebrate Paleontology* 10(4): 518–522.

Zachos, J. C., J. R. Breza, and S. W. Wise. 1992. Early Oligocene ice-sheet expansion on Antarctica: Stable isotope and sedimentological evidence from Kerguelen Plateau, southern Indian Ocean. *Geology* 20: 569–573.

Zachos, J. C., K. G. Lohmann, J. C. G. Walker, and S. W. Wise. 1993. Abrupt climate change and transient climates during the Paleogene: A marine perspective. *Journal of Geology* 101: 191-213.

Zinsmeister, W. J. 1979. Biogeographic significance of the late Mesozoic and early Tertiary molluscan faunas of Seymour Island (Antarctic Peninsula) to the final breakup of Gondwanaland. In J. Gray and A. J. Boucot, eds., *Historical Biogeography, Plate*

Tectonics and the Changing Environment, pp. 349–355. Corvallis, Oregon: Oregon State University Press.

Zinsmeister, W. J. 1982. Late Cretaceous–early Tertiary molluscan biogeography of southern circum-Pacific. *Journal of Paleontology* 56: 84–102.

Zinsmeister, W. J. 1989. A look at the dramatic change in molluscan diversity along the southern margin of the Pacific at the end of the Eocene. *Geological Society of America, Abstracts with Programs* 21(6): A88.

Index

INDEX